THE
PREPPER'S
POCKET GUIDE

THE PREPPER'S POCKET GUIDE

101 EASY THINGS YOU CAN DO TO READY YOUR HOME FOR A DISASTER

BERNIE CARR

Illustrated by **EVAN WONDOLOWSKI**

Ulysses Press

Published in the U.S. by
ULYSSES PRESS
P.O. Box 3440
Berkeley, CA 94703
www.ulyssespress.com

ISBN: 978-1-56975-929-5
Library of Congress Control Number 2011925189

Printed in Canada by Marquis Book Printing

20 19 18 17 16 15 14 13 12 11 10 9 8

Acquisitions Editor: Keith Riegert
Managing Editor: Claire Chun
Project Manager: Kelly Reed
Editor: Richard Harris
Copyeditor: Lauren Harrison
Proofreader: Barbara Schultz
Production: Judith Metzener
Illustrations: Evan Wondolowski
Front cover design: Double R Design
Back cover design: what!design @ whatweb.com
Cover photos: plastic bottle © Picsfive/shutterstock.com; duct tape
    © Feng Yu/shutterstock.com; light © Isantilli/shutterstock.com;
    first aid cross © Skovoroda/shutterstock.com; tin can lid © Matthew
    Cole/shutterstock.com; bandage © cloki/istockphoto.com

Distributed by Publishers Group West

To my family—you are the reason I prepare.

CONTENTS

Introduction...13

Chapter One: Getting Started...17

1. De-Clutter Your Home...17

2. Get Creative About Storage Space....................................18

3. Rethink Your Mind-Set..20

4. Make a Written Emergency Contact List........................22

5. Make a Family Emergency Plan...23

6. Make a Master List of Passwords......................................24

7. Assemble a 72-Hour Survival Kit.......................................25

8. Avoid Common Preparedness Mistakes..........................28

9. Carry These Important Items Daily....................................30

Chapter Two: Financial Readiness..32

10. Create a Personal Economic Disaster Plan.................33

11. Start an Emergency Savings Fund...................................34

12. Pay Off Debt with the Snowball Method......................36

13. Review Your Insurance Coverage....................................37

14. Keep an Emergency Cash Stash in Your Home............38

15. Consider Investing in Precious Metals..........................39

Chapter Three: Water Needs..44

16. Calculate How Much Water You Need............................44

17. Start Collecting Water for Emergencies.......................45

18. Find Hidden Backup Water Sources in Your Home.....47

19. Learn to Empty Your Water Heater.................................48

20. Find Water Outside Your Home..50

21. Keep Contaminated Water from Entering
Your Home...51

22. Learn to Filter Water 52

23. Make a Simple, Inexpensive Homemade
Water Filter...53

24. Learn to Purify Water...55

25. Learn to Distill Water...57

26. Learn to Build a Solar Still..58

27. Learn to Build a Vegetation Still 60

28. Learn to Disinfect Water with Sunlight 61

29. Choose a Commercial Water Filter.................................62

30. Conserve Water in an Emergency....................................63

Chapter Four: Food Supplies...64

31. Track What You Eat 65

32. Learn Basic Cooking Skills 67

33. Start a Food Storage Plan for $5 a Week 68

34. Buy What You Eat, Eat What You Store 72

35. Stockpile Food...73

36. Rotate Your Food Supply...74

37. Know the Different Types of Expiration Dates 75

38. Consider Storing MREs...76

39. Store Dehydrated and Freeze-Dried Food....................77

40. Decide How Much Food to Store......................................79

41. Buy Staples in Bulk...80

42. Overcome the Enemies of Food Storage 84

43. Learn to Sprout Seeds and Beans......................................86

44. Learn to Cook Lentils and Beans 88

45. Learn to Make Homemade Yogurt 90

46. Learn to Marinate Vegetables..92

47. Learn to Make Freezer Jam...94

48. Learn Basic Canning Methods 95

49. Try Sample Canning Recipes.............................99

50. Dry Herbs and Peppers without a Food
Dehydrator 102

51. Make Your Own Beef Jerky 103

52. Learn to Make Bread from Scratch 105

53. Start an Inexpensive Container Garden..............107

54. Start Your Own Seeds....................................108

55. Start a Garden in a Small Space 109

Chapter Five: Ready Your Home.............................**112**

56. Teach Children About Prepping 112

57. Don't Forget About the Pets............................114

58. Keep Important Documents in One Binder..............115

59. Secure Family Photos and Keepsakes 117

60. Make a Safe from a Hollowed-Out Book 118

61. Assemble a Car Emergency Kit.........................120

62. Assemble a Desk Emergency Kit 122

63. Assemble a Mini Survival Kit in a Mint Tin 123

64. Stock Up on Multipurpose Items.......................124

65. Make a Can Organizer for Your Pantry 130

66. Prepare for an Earthquake..............................131

67. Prepare for a Hurricane..................................134

68. Prepare for a Tornado....................................136

69. Prepare for a Flood 138

70. Prepare for an Ice Storm................................141

Chapter Six: Personal Health and Safety.............**143**

71. Follow the Number-One Safety Rule 144

72. Protect Yourself from Crime 145

73. Know Basic Fire Safety 148

74. Protect Your Family from Germs 149

75. Learn to Make a Disinfectant Cleaner from Emergency Kit Items 151

76. Take a Fitness Reality Check 152

77. Assemble an Inexpensive First Aid Kit 153

78. Assemble an Emergency Dental Kit 156

79. Make an Electrolyte Solution to Prevent Dehydration 157

80. Learn to Make a Gel Pack for Muscle Pain Relief 158

81. Learn to Make a Saline Spray for a Stuffy Nose 159

82. Get by with What You Have 160

83. Learn to Make a Temporary Toilet 164

84. Learn to Dispose of Waste Matter 165

Chapter Seven: When the Power Is Out 166

85. Assemble a Power Failure Kit 167

86. Know What to Do if You Have a Power Outage 168

87. Consider Buying a Generator 169

88. Learn to Store Gasoline Safely 170

89. Learn to Cook Without Electricity 172

90. Learn to Build an Outdoor Pit Oven 173

91. Know Different Ways to Start a Fire 175

92. Consider Alternative Lighting 177

93. Learn to Make a Perfect Cup of Coffee Without Electricity 179

Chapter Eight: When You Have to Get Out 184

94. Pack a Bug-Out Bag 185

95. Determine Whether You Should Stay or Go 189

96. Make an Evacuation Plan in Advance 191

97. Find Exit Routes from the City 194

98. Have the Right Footwear for Walking Out.................195

99. Keep Paracord on Hand.................................197

100. Learn to Navigate Without a Compass......................198

101. Learn to Signal for Help if You Are Stranded...........200

A Final Word 203

Resources 204

Acknowledgments...211

About the Author...213

Introduction

THERE IS A LOT OF UNCERTAINTY in our daily lives. After years of prosperity, the United States and much of the rest of the world entered the Great Recession in late 2008. Millions of American jobs were lost, and real estate prices tumbled like a house of cards. The ensuing foreclosures and bankruptcies continue to add to the ranks of the homeless.

Besides the financial crisis, on the daily news we witness natural disasters such as floods, earthquakes, tsunamis, landslides, hurricanes, tornadoes, and wildfires, as well as the man-made disasters of terrorism and war.

Like many people, I started getting the feeling of insecurity and unease on a daily basis. I felt worried about myself, my family, and the future. Yet I found myself at a loss as to what to do about it.

I needed to regain the feeling of security and hope, so I started to delve into the world of preparedness. I read government pamphlets and websites about emergency preparation and researched ways to survive. I found a lot of good information, but it was not always geared toward beginners like me, who live in the big city in apartments or small houses with no extra space, who must live within tight budgets, who do not have access to farmland or other faraway retreats, who have families and pets, and who hold full-time jobs and do not have a lot of extra free time.

As I read, I adapted what I learned to my individual situation. Now I know that I am not alone in my concerns by any means and that there are simple ways to prepare our homes and our families, one step at a time.

When we start feeling uncertain and insecure about our situation, we feel isolated and alone. We think our feelings are unique and no one would share our beliefs. To a certain extent, the majority of the population lives day-to-day without thought about what might happen should the system fail. Yet I've found that a lot of like-minded individuals are also concerned about their families and do not want to rely on the government or the system in case of emergency, but they don't know where to start. I wrote this book to share information with those people.

Why Should We Prepare?

Most people have some form of insurance on their cars, health, homes, and lives. We pay premiums on a regular basis just in case something happens and we end up needing the coverage. No one hopes to get sick or have an accident, but we pay the fees anyway. In the same way, no one hopes for an emergency or disaster to happen, but we ought to prepare for one and in that way build a sense of security and control over our surroundings.

Myths about Preparing

There are several notions about preparing that turn out to be myths. The very mention of survival or preparedness can conjure negative mental images that have no basis in fact.

MYTH 1: Preparing Is Expensive   Preparing does not need to cost a lot. You can start small and only purchase things within your budget. In fact, some preparations, such as food storage, can actually save you money.

MYTH 2: Preparing Takes Too Much Time   Anything, including hobbies, surfing the web, or watching TV, can take "too much time." Preparing is a worthwhile activity to which you can devote as much or as little time as you want. It all depends on your comfort level.

MYTH 3: You Need a Lot of Space for Storage   Anyone can create storage space, whether in a small apartment or a house in the suburbs.

MYTH 4: You Need a Farm or a Retreat Location   It would be nice if we could each lay claim to a faraway place we could run to, but most of us are not that fortunate. Instead, just prepare wherever you are, as best as you can. Any amount of preparation is better than none at all.

MYTH 5: Preparing Will Turn Me Into One of Those Crackpots Living in a Cabin in the Woods, Dressed in Military Gear and Threatening People with Explosives   This image stems from movie stereotypes of survivalists, sociopathic loners like "Unabomber" Ted Kaczynski and right-wing militias that thrive on visions of far-fetched conspiracies to justify firearm fanaticism. The truth is, most "preppers" (a term for the new breed of survivalists now widely used by sources from *Newsweek* to Wikipedia) are normal everyday people who realize the system may not always be around to support them and so feel they need to provide for their own safety and security. In an era of ongoing financial crises, spectacular natural disasters, and technology run amok, imaginary conspiracies by shadowy government entities are really the least of our problems. So let's get real, shall we?

There are a lot of emergencies we can prepare for, ranging from events with a higher likelihood of occurring, such as unemployment or illness in the family, to natural disasters such as hurricanes or earthquakes to rare but extremely catastrophic events like a terrorist attack or pandemic. Whether you're making preparations for a short-term or long-term emergency, the initial steps toward those goals are the same.

Even if you are starting from the beginning, with no emergency supplies at all, once you get started with the prepping process, you will soon start seeing positive results. As you build on these simple steps, you will become more confident in your efforts and be rewarded with the confidence that you have done the best you can to protect your family from disaster.

CHAPTER ONE
Getting Started

WHEN I FIRST STARTED TO PREPARE, I felt overwhelmed by the sheer amount of information there is to learn. I'm here to tell you it doesn't have to be complicated. You can start out slowly and work at your own pace. How much and how quickly you gather supplies or learn skills is purely up to you. There are no deadlines, and no one is going to judge you except yourself. You will be able to gauge your efforts by your own feelings about how secure you are starting to feel.

1

DE-CLUTTER YOUR HOME

You will need to free up some space to store your emergency preparedness supplies. An unused room or closet is a good place to start storing your inventory. Make sure the area is clean and easily accessible. If you live in a limited space—an apartment or small house—you have an even more urgent need to pare down and de-clutter your space.

Before starting on your preparedness program, take a look around your home. Inventory what you have. How much food do you have in your pantry and refrigerator? How much water do you have in your house right now? What do you have stored in your closets? Are these spaces organized so you can easily find what you need?

Consider each item and how often you have used it. If you have items in the pantry and fridge that are expired, it's time to toss them out. Check your closets and other storage areas and ask yourself which items you no longer use. Do you have books you have already read and probably won't read again? Movie DVDs you likely won't watch again? Keep in mind that even if you might want to read or watch them someday, that's what libraries and video rental stores are for. You can even turn them into cash by selling them to a used bookstore or video store, or donate them to charity and take the tax deduction. If you have any items that have not been used in more than a year, then it is time to consider getting rid of them.

2

GET CREATIVE ABOUT STORAGE SPACE

After you have de-cluttered as much as possible, you may find that you still need to create more storage space. Keep track of all your hiding places by making an alphabetical master list of where everything is stored. This way, if you have to rush out of the house in an emergency, you can gather up everything quickly. Here are a few ideas to consider:

BEDROOM AND CLOSET   Store smaller items inside larger items. For example, empty suitcases can be used to store survival supplies. Use the space under the beds. Items that can be flattened, such as blankets and comforters, can be stored in plastic bags and stored under the bed. For more space underneath, use bed risers to elevate the bed.

BATHROOM   Bathrooms often have places where you can install more shelves. Add stand-alone vertical shelves in empty corners. Build shelves over the toilet.

LAUNDRY ROOM   Build horizontal shelves over the washer and dryer.

LIVING ROOM AND DINING ROOM   An entertainment cabinet can be repurposed to store emergency supplies if it has extra cabinets that are not enclosed with glass. A long tablecloth on the dining table or any other surface can be used to hide a few boxes underneath, as long as you leave some foot room. A wooden chest can serve a double purpose as a coffee table with storage space inside. No one needs to know you have canned goods in there! A window bench that has storage space under a hinged seat can store emergency supplies such as a lighting or blackout kit.

GARAGE   Build more shelves.

3

RETHINK YOUR MIND-SET

OVERCOME THE NORMALCY BIAS AND SAVE YOUR LIFE Normalcy bias is a state of denial that many people get into when faced with an impending disaster. They underestimate the seriousness, as well as the aftereffects, of the disaster that is happening all around them. They become immobilized and slip into a "deer in the headlights" paralysis. Unfortunately, this common reaction is worsened by stress. It is human nature to assume that because nothing huge and dangerous has ever happened around here before, no such catastrophe will ever happen. For example, when Hurricane Katrina was about to hit New Orleans, despite warnings to evacuate, many people chose to stay in their homes.

Avoid being caught unprepared by thinking ahead about possible emergencies. Make a conscious choice to overcome the effects of normalcy bias, and you will be more mentally prepared to cope with an impending disaster. Be aware of what is going on around you; realize and accept the possibility of an emergency. Learn to recognize threats. Never assume that things will go along as they always have.

First and foremost, consider the possibilities of various types of emergencies and make a plan for your family's safety.

TAKE RESPONSIBILITY FOR YOUR OWN SURVIVAL Do not rely on anyone to save you.

The sooner you accept responsibility for your own survival, the safer you will be. During an early stage of a disaster, people take on a calm and courteous demeanor, as though they do not want to be the first to bail out. They might mill around, waiting for someone to take charge and tell them what to do. Unfortunately, waiting might cost them their lives. Once you recognize impending danger, get yourself to safety as soon as possible.

When faced with a disaster, remember the "STOP" rule:

STOP — Take a deep breath and recognize what is happening around you.

THINK — Don't panic; think through your predicament before you react.

OBSERVE — Look at what's going on and assess your situation.

PLAN — Think about how you are going to deal with the emergency. Then follow through.

4

MAKE A WRITTEN
EMERGENCY CONTACT LIST

Most of us keep our important names, addresses, and phone numbers in our cell phones or computers, but it is a good idea to have a hard-copy backup in case you lose your phone, service is interrupted, your computer gets a virus, your hard disk crashes, you run out of batteries, or there's a power outage. Print your entire contact list and keep it in a safe place. Reprint it whenever you add to it or make changes. Another option is to copy your contacts into an old-fashioned paper address book

Post an emergency contact list with addresses and phone numbers in a visible place in your home, such as a refrigerator door or bulletin board. Include phone numbers for your nearest relative or close friend, out-of-state family or friends, family physician, pediatrician, poison control, dentist, and utility providers, including water, power, gas, cell phone, cable, and Internet.

5

MAKE A FAMILY EMERGENCY PLAN

In the aftermath of the 2011 earthquake in Christchurch, New Zealand, family members spent hours trying to find each other, as the earthquake happened in the middle of the day when people were at work. Many people were unable to find their loved ones for several hours. To avoid this kind of personal chaos after an emergency, plan ahead and decide how your family members will get in touch with each other or where you will go in case of an evacuation.

For example, on weekdays the parents may be at work and the children will be at school. If something happens in the middle of the day, who will pick up each child? Will Dad or Mom pick up all the kids and meet at home? Or will both spouses share the responsibility? If the emergency makes your home unsafe and you must evacuate, do you have a backup place to meet?

Make an emergency card that each family member can carry around, including young children. The list below includes suggested information, or you can tailor the information to suit your own family:

Emergency Card
❑ Mom's cell phone, work phone
❑ Dad's cell phone, work phone

- ❏ Home phone
- ❏ Neighborhood meeting place
- ❏ Out of town meeting place
- ❏ Out of town contact

Have each child's school schedule handy so the parents know where each child is at any given time. Know your children's school's emergency plans, as well as your workplace's emergency plans.

6

MAKE A MASTER LIST OF PASSWORDS

We have so many passwords and personal identification numbers (PINs) for work, shopping websites, banking institutions, and so on that it is easy to lose track of them. Keep backup lists of all your passwords, one for work and one for personal. Keep an electronic backup of the password list protected by—you guessed it—a password; keep another copy of the list on a piece of paper concealed in a safe hiding place. When you write your passwords down, do not use the real account names; use code words only you would know. For security purposes, do not use the same password for all your accounts.

7

ASSEMBLE A 72-HOUR SURVIVAL KIT

Remember the survival "rule of three": A person can survive three minutes without air, three hours without shelter, three days without water, and three weeks without food. Of course, other factors may affect your chances for survival, such as your health, your mental state, and available resources. A 72-hour survival kit, the cornerstone of your preparedness efforts, will help you and your family survive in case of emergency if outside help does not arrive right away. After you have the 72-hour kit in place, build on it and add more survival supplies until you have at least two weeks' worth.

WATER Water is essential to life, so always have three days' worth of clean water on hand. Plan on one gallon per person per day for drinking, cooking, and basic sanitation such as brushing teeth. You'll also want a way to filter and disinfect more water.

FOOD You need food to survive and give you energy. Food is the most perishable of the basic necessities, so you'll want to stock canned or freeze-dried foods in case refrigeration is unavailable. The food should not require gas or electricity to prepare, in case the power is out, although a propane camping stove is a blessing, especially for preparing freeze-dried foods. Include foods that your family likes to eat—perhaps canned meats, fruit and vegetables, fruit cups, granola bars, and dried pasta. Don't forget to include comfort items such as snacks, coffee or tea, candy, and chocolate, as well as special items such as baby formula to suit your family's particular needs. If you have pets, make sure you have enough food for them as well. Include a can opener in your food supply kit, plus paper plates and cups to avoid having to wash dishes.

You'll need large heavy-duty trash bags that seal well to keep household garbage until trucks are able to collect it.

FIRE AND LIGHT   Fire provides warmth and comfort as well as a way to cook food and disinfect water. Be sure to keep a supply of firewood, paper, kindling, and a waterproof container of matches. Emergency candles, lanterns, and flashlights with plenty of extra batteries are also essential when the electricity is out.

SHELTER AND CLOTHING   A tent or tarp offers protection from the elements, and clothes help regulate your body temperature. Keep a supply of clothes you can wear in layers for maximum adaptability, including a waterproof outer layer. Keep in mind that cotton clothing loses its insulating ability when wet. Launder your clothing before storing it; it will get dirty soon enough if the time comes when you need to use it.

COMMUNICATIONS   A cell phone can be a lifesaver for helping search-and-rescue teams to locate you after a disaster. Keep your regular cell phone with you, because the battery will eventually run down in a phone left unused in your emergency kit for weeks or months, and leaving a cell phone plugged in for long periods of time will also shorten battery life. You can store one or more spare phone batteries in your kit, though. A battery-powered or hand-crank radio can help you stay informed about conditions beyond your immediate area.

HEALTH AND PERSONAL HYGIENE   A first aid kit is essential and should include basic wound care such as bandages and antibiotic ointment, pain relievers, and diarrhea medicine, along with any allergy medicine and personal prescriptions your family may require. You'll also want to keep extra eyeglasses or contact lenses on hand. Hygiene supplies should include toothbrushes, toothpaste, soap, hand sanitizer, toilet paper and paper towels, sanitary napkins or tampons, and bleach or disinfectant. If you have an infant in the family, of course you'll want diapers—preferably reusable cloth ones—and baby wipes. And if you have cats, don't forget the kitty litter. A portable camping toilet can offer a more civilized alternative to a latrine ditch behind nearby bushes.

OTHER TOOLS   Some of the most useful tools are a utility knife, duct tape, gloves, pens, and paper.

ENTERTAINMENT   During those long nights when you're stranded by a disaster and there's no electricity, iPads and portable mp3 or DVD players are all very nice to have, but after the batteries run down you'll be glad you thought to store board games, a deck of cards, and a selection of paperback books.

SECURITY   Protecting yourself and your family from harm is always a consideration. What that means may depend on your personal philosophy. Some people wouldn't think of leaving the house without a firearm, a Taser, or at least pepper spray. Others worry that the potential harm from keeping weapons around far outweighs possible self-defense benefits. The choice is yours.

EMERGENCY CASH   If you're forced to evacuate your home, you'll need cash. If you're stranded on your roof during a flood, it may be superfluous. But whatever your circumstances, it can't hurt to have money on hand.

8

AVOID COMMON PREPAREDNESS MISTAKES

IMPROPER FOOD STORAGE   If you are going to store food long-term, learn the proper food storage techniques (see page 82) before embarking on your project. Much stored food can go to waste due to pests, moisture, heat, and other factors that can degrade food.

NOT ENOUGH FOOD VARIETY   Store a variety of foods for a well-balanced and varied diet. Eating rice every day may fill you up, but eventually you will get sick of the "same old same old." Include fun foods such as hard candy, chips, and chocolate that can give you a psychological boost in a difficult situation.

FORGETTING ABOUT YOUR STORED FOOD Don't store food for an indefinite period without rotating it. Check expiration dates regularly so you will be reminded to use the food before it goes bad. Learn how to cook with your various stored foods, whether they are canned, dehydrated, or freeze-dried.

OVERSPENDING ON GEAR There is so much equipment and supplies available that it's easy to overbuy. Don't spend your money all at once on the latest gadgets. When shopping, ask yourself how essential a combination GPS and corkscrew would really be in the event a tornado hits your neighborhood. Wouldn't you rather have the cash it costs? Assess your needs first, and then shop carefully, choosing multipurpose items and comparing prices.

NOT ACQUIRING SKILLS Being prepared requires more than just stuff. Learn new survival skills to become more self-sufficient.

WORRY AND PANIC There is always a lot of doom and gloom talk around. If you listen to all the dire predictions, you can become paralyzed with fear. The secret is to just do a little bit at a time. Buy supplies and acquire skills, but live within your means and make your preparations one step at a time. The feeling that you are more prepared than you were before will soon assuage worries.

COMPARING YOURSELF TO OTHERS   When you first start preparing, you may feel like there's so much to do and others are way ahead of you. Sure, there will always be people who got going earlier, but they had to start in the same place you are right now. As long as you get started and go at your own pace, you will feel more and more secure.

CARRY THESE IMPORTANT ITEMS DAILY

We heard about so many victims in the March 2011 Japan earthquake who got caught in the disaster while far from home. Most people have some emergency supplies at home, and prepping your house and family is important, but a disaster can happen anywhere at any time, so it is a good idea to keep certain preparedness items with you at all times. You do not need to carry a large amount of stuff, just a few choice items that can help until you get to safety. Most likely you already have some of them.

Here are a few suggestions:
- ❏ Wallet with your identification, insurance information, and debit and credit cards
- ❏ Cell phone
- ❏ At least $20 to $40 in cash
- ❏ Small flashlight

- ❑ Small bottle of water (8 ounces)
- ❑ Granola bar or trail mix, and a few pieces of hard candy
- ❑ Sunglasses
- ❑ Pen
- ❑ Small wallet for first aid items like pain reliever, diarrhea tablets, and allergy medicine
- ❑ Safety pin
- ❑ Multi-tool like a Swiss Army knife (*to be carried only if you are not catching a flight*)
- ❑ Pepper spray (if allowed in your state; *to be carried only if you are not catching a flight*)

CHAPTER TWO

Financial Readiness

WHEN WE THINK ABOUT DISASTERS, it's usually earthquakes, floods, fires, and even terrorist attacks that come to mind. But by far, the most common disasters are financial in nature. Sudden unemployment or catastrophic medical expenses can happen to anyone, and of course, cataclysms like floods and fires also have severe financial consequences.

I was once laid off twice in the same year. Soon after the first time I lost my job, I got laid off a second time by a different employer. Unfortunately, I had already spent my severance check from the first job, and I got no severance funds whatsoever from the second job. During the recession, many people have found themselves in similar situations. In uncertain economic times, it is even more imperative to prepare yourself financially as quickly as possible.

10

CREATE A PERSONAL ECONOMIC DISASTER PLAN

To truly prepare yourself, you need to plan what you would do if you had your own personal economic disaster. This is not a matter of pessimistically expecting the worst; rather, it is reasonable, well-thought-out preparedness.

Imagine for a moment what would truly happen if you lost your job, whether you are single or have a family to provide for. What if you and your spouse were to lose your jobs at the same time?

Given your present resources, how many months could you continue paying bills with no monthly income? Sure, you can apply for unemployment benefits, but what if your ex-employer contests your claim and your application is rejected? What if you do not qualify for unemployment because you haven't worked at that job long enough or your employer goes out of business?

Consider how many months of rent and utilities you would be able to cover with your savings. What bills would you continue to pay? Rent or mortgage would be the first priority, because you need a roof over your head. Food and utilities would come next, along with car payments for at least your main car, which you'll need get around and find a new job. Other discretionary spending, such as cable TV, eating out, and entertainment should be slashed. Everything else would come after. In the meantime, until you found a new full-time job, a combination of part-time jobs might be an option.

You might discuss lower monthly payments with your other creditors. If savings were to run out, you would try to negotiate with your mortgage holder or landlord to try to stretch out your payments. You might even have to move in with family members for a short time until you get back on your feet.

Would you move to another state if jobs were scarce where you live?

These are just a few considerations to work through when preparing for a personal financial crisis. Having a plan in place will give you peace of mind and at the same time help you pinpoint your vulnerabilities so you can improve your current situation.

11

START AN EMERGENCY SAVINGS FUND

Start saving money so you have at least three months of living expenses—many financial advisors say six months. You should have enough savings to cover your rent or mortgage and basic utilities such as electricity and water, food and transportation. If you already feel tight financially, think how much worse it could be if you lost your job and found yourself unable to receive unemployment benefits. Here are a few tips to get your emergency fund going:

PAY YOURSELF FIRST Before paying bills, going out to celebrate on a Friday night, or anything else, set aside some savings—even if you can only set aside $10 per week. Save the money in a hidden jar at first. Once you have $50 or more, start a free savings account at your local bank or credit union. Make sure there are no fees involved. Then set up a direct deposit from your checking account to your savings, $25 at the minimum, and you will not miss the money.

One good way to get your savings started is to de-clutter your home and sell unwanted items. Have a garage sale, or sell your stuff on eBay or Craigslist and set aside the money you make in your emergency savings fund.

LOWER YOUR BILLS Look at every monthly bill you pay and evaluate whether you can do without it. Can you switch to a lower cable, Internet, or cell phone plan? Call your carriers and find out what your options are for switching plans or negotiating lower rates. Competition between cell phone companies is stiff, so often you will qualify for a lower rate after you've been with the same company beyond the minimum contract period—but often they won't tell you this unless you ask. When you succeed in lowering expenses, immediately send the difference to your savings account.

12

PAY OFF DEBT WITH THE SNOWBALL METHOD

If you have credit card debt, stop using your credit cards now. Start living on a cash-only basis and put away your credit cards in a safe place.

Make a plan to pay off all your consumer debts. First make a list of all the amounts that you owe and the minimum payments for each one. Apply any extra money that you have toward the lowest-balance card. This will give you the immediate gratification of paying off a debt quickly. Pay the minimum to the rest. As you pay off one card, add the amount you were paying on it to the payment you're making on the next card. Some experts recommend a different strategy of paying off the highest annual percentage rate card first to maximize your interest savings. The trouble with that plan is that if it's not your lowest-balance card, it may take you longer to pay off. Paying off the low-balance card first, you will feel encouraged by a sense of progress. Also, if your cards carry annual fees, lessening those extra charges by reducing the number of cards you have may save more than cutting the interest expenses.

Call your credit card company and see if you can negotiate a lower interest rate, especially if you have good credit and are receiving other solicitations to transfer your balance.

13

REVIEW YOUR INSURANCE COVERAGE

Make sure you have health- or medical-insurance coverage for everyone in your family in case of any emergencies.

Be sure you have life-insurance coverage for yourself and your spouse, especially if you have young children who depend on you. Uninsured major medical expenses are the biggest single reason for bankruptcies in the United States.

If you own your home, make sure you have homeowners insurance in force. Most mortgage lenders require it, but some will let you waive it if the value of the land exceeds the amount of the mortgage. Not all homeowners policies offer the same coverage. Be sure you're covered not only for damage to the home but also loss or theft of its contents, additional living expenses if it becomes uninhabitable, and liability for accidents on the premises. In areas where floods are common, most homeowners insurance policies exclude flood damage, so you'll want supplemental insurance to cover that possibility; the same is also true of other natural disasters such as fires or earthquakes. If a disaster happens, insurance is essential to help you recover. If you are a renter, consider rental insurance for your house or apartment.

14

KEEP AN EMERGENCY CASH STASH IN YOUR HOME

Many people no longer use cash, instead preferring debit cards for everyday use. But you should keep some emergency cash in your home in case you cannot access the money in your bank account. Recently, a computer glitch caused thousands of people in Australia to lose access to their bank accounts for several days. During an emergency, the power could go down, rendering ATMs useless. To make sure you have sufficient funds for everyday needs, such as food and fuel, keep cash in a secret hiding place. Here are a few ways to help build up a small household cash fund for emergencies.

AT THE STORE Keep all "found money" in a change jar. Even if you find a penny on the ground, save it in the jar. Every now and then, convert it to paper currency at your supermarket's coin-sorting machine. Use coupons at the supermarket. Then, from the change you receive, set your coupon savings in cash. Sign up for product rebates. When you receive the check, cash it and save it in your cash stash.

SET ASIDE SAVINGS ON UTILITIES Conserve water and electricity. When you get a bill that is lower than your budgeted amount, even if it's just a few cents, take the difference out of your checking account and put it in your cash stash. Call your cable, phone, and cell phone providers and ask the representative to help you go over your bill. Review each service you are paying for, and ask for help in trying to lower your bill. You might consider moving to a lower-cost plan. Set aside your savings in the cash fund.

15

CONSIDER INVESTING IN PRECIOUS METALS

Gold and silver can be a part of your overall financial portfolio, but only after you've acquired a good amount of emergency supplies, created your emergency savings and cash funds, and paid off some debt. In some of the world's less stable nations, they are a primary form of savings and investment, not only providing a hedge against inflation but also protecting against currency becoming worthless after a regime change or a bank failure, and providing a way to carry wealth across international borders. In the United States, many economic fearmongers tout gold and silver as protection against "the coming financial collapse." But there are pitfalls to be aware of.

Should you run out and buy gold or silver before other supplies? Or go into debt so you can have precious metals? No. Remember, no new debt. You should first make sure you have your emergency water, food, and other survival supplies, then set aside your emergency funds and pay off debt. In an emergency, if you can't access your bank account or the stores are closed, you are better off having water and food stored to feed your family than precious metals close at hand.

Buying Precious Metals

If the time comes when you are ready to buy precious metals, keep these tips in mind:

FIND A REPUTABLE GOLD DEALER It is best to find a source such as a rare coin dealer, preferably recommended by someone you trust. Independent jewelers and dental laboratories have accounts with gold suppliers and are often willing to buy and sell gold on your behalf for a commission. Fully research and investigate the dealer, and check ratings online and complaints from the Better Business Bureau.

CALL AHEAD Check in advance to inquire what they have available on the day you plan to stop by. Especially in the coin market, inventory may vary day by day.

KNOW THE VALUE OF YOUR METALS Precious metals have both "intrinsic" and "collectible" value. Intrinsic value is the weight of the bullion or coin multiplied by the spot price, which fluctuates constantly. Collectible value is the price of the metal coins or jewelry above the intrinsic value. When liquidating gold, you can count on getting most of its intrinsic value for it, minus a brokerage commission, but the collectible value is likely to be two or even three times as much when you buy it as when you sell it. You get more gold for your money if you buy bullion, because the markup is not as high as it is for collectible coins. But bullion comes in larger weights—typically 10 troy ounces to one kilogram (32.15 troy ounces). Because the price of gold at this writing is between $1,400 and $1,500 an ounce, even a small gold bar represents a substantial investment on the same order as the price of a good used car.

Gold coins and jewelry might not hold their value as well, but if you were ever in a situation where you needed coins for barter, small-denomination coins would come in handy. Some gold coins are minted by governments and were originally used as currency, though they are mainly made for collectors today. These include the famous South African Krugerrand, the original government-minted bullion coin, as well as the gold American Eagle, Canadian Maple Leaf, the Chinese Panda, the Mexican Centenario, and the Austrian Corona. There are also many gold and silver coins produced by private mints, but they often sell at higher collectible markups and are harder to sell for anything close to what you paid for them.

TRACK DOWN SILVER COINS One way to start acquiring silver is to buy pre-1965 coins such as the Kennedy half dollar and the Peace dollar. Coins minted in the U.S. since that time have less silver content—and sometimes none at all—though they are still valued by collectors.

Make sure you are buying actual gold or silver that you can take possession of, not a certificate for future delivery or investor funds backed by metals.

Selling Your Precious Metals to Raise Cash

Just as when you buy precious metals, make sure you do business with a reputable dealer. Meet them at their store or office; do not mail your gold to an unknown party or arrange to meet anyone at your home or in a stranger's place. Typically you can sell your gold to a rare coin dealer, a jeweler who buys gold, or a reputable pawnshop. For gold, in most states, you will need a photo ID, and you may need to sign an affidavit stating where you got the gold.

Know the amount of gold you are taking with you, and check online for the current spot price of gold so you'll know enough to negotiate. Gold is measured in troy ounces (which are about 10 percent heavier than regular ounces). The gold weight may be less than the full weight because most gold used for manufacturing is alloyed with other metals to make it harder. Modern gold coins usually contain copper, and gold jewelry contains copper and either silver (for yellow gold) or nickel (for white gold). The purity of gold in coins or jewelry is measured in carats (k): 10k, 14k, 18k, or 24k are most common, 24-karat being pure gold.

Gold bullion's purity is measured in percentages, typically .995 fine or .9999 ("four nines") fine. Dealers will not buy gold-filled jewelry.

If you are selling silver, be aware of the silver content in your coins or silverware. Again, most collectible silver is alloyed with copper or other metals. The standard, sterling silver, is .925, or 92.5 percent, pure silver, while some jewelry is .995 fine silver. The value of gems set in your jewelry is much less predictable, unless you have a certificate from a registered gemologist. If the dealer will not pay a fair price for the gems, go somewhere else.

CHAPTER THREE
Water Needs

STORING WATER IS CRITICAL to every emergency plan. According to the survival "rule of three," you cannot survive more than three days without water. During many natural emergencies, such as hurricanes or floods, the water supply may be unsafe to drink. If a catastrophe cuts off pure water sources for longer than a day, waterborne diseases are a serious risk. Water is easy to set aside. But first you should calculate your family's water needs.

16

CALCULATE HOW MUCH WATER YOU NEED

The guideline is one gallon per person, per day, for drinking. That's the minimum. During an emergency you will also need water for washing your hands, brushing your teeth, washing dishes if you don't have disposable plates, washing fruit and vegetables, and other sanitation needs. For example, a 72-hour emergency kit for a family of four would need 24 gallons of water for drinking only, but more would certainly be welcome. The basic 24 gallons of water will weigh about 200 pounds and will require 3.2 cubic feet or more of storage space.

17

START COLLECTING WATER FOR EMERGENCIES

BOTTLED WATER Every time you go grocery shopping, pick up one or two plastic gallon jugs of drinking water until you have the minimum amount for drinking of one gallon per person per day. Even if you buy only two extra gallons a week, you will quickly build up your emergency water.

COLLECTING WATER IN CONTAINERS If you want to avoid the extra expense of store-bought water, start collecting clean two-liter soda bottles. Two liters equals about .528 of a gallon, so two large soda bottles per person will be enough for drinking for one person for one day. Refilling milk jugs is not recommended, as they may contain bacteria that will spoil your water. Besides, they tend to smell after a while no matter how well you clean them.

To clean and disinfect your soda bottles, first, clean out the bottles by rinsing them with tap water. To sanitize, mix one teaspoon of unscented chlorine bleach with four cups of water in the bottle. Cover with the cap and shake the mixture so that the entire inside surface is wet, then pour the mixture out. Rinse both the bottle and the cap thoroughly with clean tap water. Fill with clean tap water and cover. Label the bottle with the date you collected it. Any lingering chlorine odor or taste should dissipate after a day or less.

COLLECTING RAINWATER   Collecting rainwater is also an option, but some states prohibit it; check the laws in your state before using this method. Purchase 15-gallon containers or 55-gallon barrels that can be left outside for the next rainstorm. They will fill much faster if you place them below drain spouts but may also catch more dirt and debris from the roof and gutters. Be sure to cover the open tops of the rain barrels with screen or wire mesh—they can pose a serious drowning hazard for animals and children. Store the water in a cool, dry area. After a few months, stored rainwater can develop moss. In an emergency, if you notice a greenish tinge in your stored water, you may still use that water for rinsing and other sanitation needs. Rainwater can be tainted by airborne contaminants ranging from dust to bird droppings to radioactive fallout, so purify it before you even think about using it for drinking.

18

FIND HIDDEN BACKUP WATER SOURCES IN YOUR HOME

REFRIGERATOR ICE MAKER Many refrigerators have automatic ice makers that produce a full canister of ice. When power goes out, collect water from melted ice—it's safe to drink.

TOILET TANK Another source of water during an emergency is the toilet reservoir tank. Not the toilet bowl (which is dirty), but the reservoir inside the unit behind you when you sit down, which contains clean water that can be used. Make sure the toilet tank does not contain a cleaning agent such bleach, ammonia, or a bowl-cleaner cake. If you feel doubtful about drinking toilet water, use it for washing or sanitation—or distill it before drinking as a last resort.

POOL OR HOT TUB Pool or hot-tub water should not be used for drinking but can be used for washing and sanitation needs. The only way to purify pool water so it's safe for drinking is to distill it. Distillation is achieved by boiling water to produce water vapor, which leaves most contaminants behind and turns back into water that is safe for drinking.

19

LEARN TO EMPTY YOUR WATER HEATER

Your water heater contains clean water that is safe for drinking. Most water heaters contain about 30 gallons of water, enough for the drinking needs of three people for 10 days. It is a good idea to learn how to drain your home's water heater in case of emergency.

You will need:
- ❑ Clean garden hose with a "female" fitting on one end
- ❑ Flat head screwdriver
- ❑ Clean containers for the water
- ❑ Gloves—use to protect your hands in case the water heater is hot

Water heaters can be either gas or electric. Turn off the power to the heating element. If it is a gas heater, the pilot light will go off.

If the city's tap water is contaminated, you don't want to let it into your water heater. Turn off the cold water supply going into the heater. You can do this by shutting off the valve so water from the outside does not go into the water heater. It's generally on a copper pipe running to the top of the tank. You can also turn off the main water supply coming into your house. Wait a few hours to make sure the water heater has cooled before draining it.

Take the hose, and find the drain valve in the bottom of the unit. Attach one end of the hose to the drain valve and place the other end over the container. Open the valve with the screwdriver. If nothing comes out right away, there may be some vacuum pressure. Open up the vacuum on top to allow air to flow in; close it back up when the water is flowing freely out of the hose. Fill up your containers and shut off the drain valve when you are done.

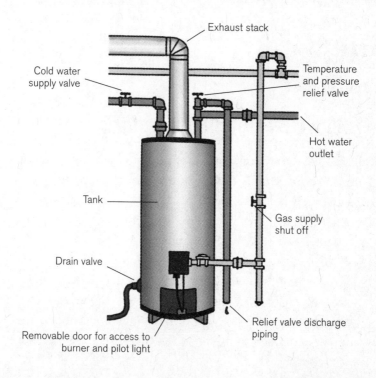

Exhaust stack

Cold water
supply valve

Temperature
and pressure
relief valve

Hot water
outlet

Tank

Gas supply
shut off

Drain valve

Relief valve discharge
piping

Removable door for access to
burner and pilot light

When the emergency has passed and services are restored, remember to turn your cold water supply back on (as well as your main water supply, if you shut it off) and let the tank fill back up. If the gas has been turned off, a gas company professional should be called in to turn it back on.

Warning: Draining the water heater should be done only in an emergency when you have no other water sources. Make sure you read the manufacturer's instructions on your tank—each water heater may vary. It is possible on some water heaters for the heating element to burn out if there is no water in the tank. While draining the water, wear gloves to avoid burning or scalding.

20

FIND WATER OUTSIDE YOUR HOME

If you have to search for water outside your home, try to find a source that is flowing rather than stagnant. Avoid water that is murky or has algae. "Lakes" in subdivisions are not safe for drinking; they are full of chemicals from lawn pesticides and other building materials.

Never drink floodwater. You can only drink saltwater if you distill it first. Drinking unclean water can be dangerous, so always heed your city's warnings whenever a "boil water" notice is in place. Unpurified water may contain any number of dangerous organisms such as E. coli, salmonella, and giardia. Drinking contaminated water can result in vomiting, diarrhea, fever, or stomach cramps, which could become life-threatening during an extended emergency. Drinking from an unsafe water source can also cause serious diseases such as dysentery, cholera, and typhoid.

If you are not sure about water safety in an emergency situation, always purify your water. The time to learn how to do it is now, while you have some time to become familiar with the process. In an extreme emergency, when the choices are limited, you will need to filter and purify as best as you can.

21

KEEP CONTAMINATED WATER FROM ENTERING YOUR HOME

If you hear news reports that there is a broken water or sewage line in your community, you can keep the contaminated water out of your home by shutting off incoming water. First, find out where your main water supply line is located. It's usually the valve leading from the outside, connected to the water meter. Turn the valve on the supply line off to keep outside water from entering your home. Make sure all adult family members know how to do this procedure.

22

LEARN TO FILTER WATER

If the water has dust and dirt, the first step is to filter out the solid particles.

You will need:

- ❏ Materials to filter the water—coffee filters, paper towels, cheesecloth, or a clean T-shirt or some other clean article of clothing
- ❏ Funnel
- ❏ Strainer
- ❏ Clean container such as clean soda bottle

Set up your filtering material over the strainer, and place it above the funnel. Place the funnel on the mouth of the clean container and pour the water over it. The resulting water will be clearer than what you started with, as the dirt and solid particles will have been filtered out. If there is still a lot of debris or sediment, repeat the process.

23

MAKE A SIMPLE, INEXPENSIVE WATER FILTER

You will need:
- ❏ Two-liter soda bottle
- ❏ Clean cotton or foam batting
- ❏ Activated charcoal granules
- ❏ Fine and coarse gravel
- ❏ Fine- and coarse-grain sand
- ❏ Coffee filter
- ❏ Mug or other container

Cut off and discard the bottom of the soda bottle. Invert the bottle so the lip is facing down. Position it over a mug or other clean container. Place the cotton batting as the first layer in the inverted bottle. Next, layer the activated charcoal, then the fine-grain sand followed by the coarse-grain sand, then the fine gravel followed by the coarse gravel. Start the layers over and repeat in the same sequence until you get to the top of the bottle. Now top the filtration system with a coffee filter. Rest the bottom of the filtration system in a mug or other wide-mouth container that will catch the filtered water. Pour water through the coffee filter. It will work through the various layers of the filtration system, leaving all the impurities behind. The impurities accumulate as sediment in the layers, and pure water flows through the straw spout into the jar. You will need to replace the coffee filter and the charcoal as they become dirty.

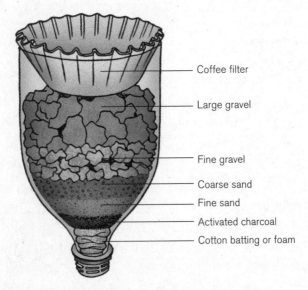

- Coffee filter
- Large gravel
- Fine gravel
- Coarse sand
- Fine sand
- Activated charcoal
- Cotton batting or foam

A homemade water filter

24

LEARN TO PURIFY WATER

After you have filtered out dirt and solids, the next step is to purify the water, eliminating any bacteria or other organisms that can make you sick. There are several methods you can use to purify the water for drinking:

BOILING WATER All you need is a clean container for the water, plus fire or another heat source. Let the water reach the boiling point. Continue to boil it for at least one minute, then remove from the heat. Wait for the water to cool before drinking it. Be aware that boiling alone will not protect against giardia, a common parasite from animal feces in streams and ponds.

DISINFECTING WATER WITH BLEACH Another method for purifying water is to use plain chlorine bleach. Generic is okay, as long as there are no ingredients other than bleach— typically 5- to 6-percent chlorine. Make sure the bleach has no perfumes, dyes, or other additives that will contaminate the water even as you purify it.

You will need:
- ❏ Chlorine bleach
- ❏ Clean container
- ❏ Medicine dropper or teaspoon
- ❏ Water to be purified

This chart tells how much bleach to use:

Treating Water with Liquid Chlorine Bleach*		
Volume of Water to Be Treated	Treating Clear/Cloudy Water: Bleach Solution to Add	Treating Cloudy, Very Cold, or Surface Water: Bleach Solution to Add
1 quart (1 liter)	3 drops	5 drops
½ gallon (2 liters)	5 drops	10 drops
1 gallon (3¾ liters)	⅛ teaspoon (8 drops)	¼ teaspoon (16 drops)
5 gallons (19 liters)	½ teaspoon	1 teaspoon
10 gallons (38 liters)	1 teaspoon	2 teaspoons

*Source: Dept of Health Pub 821-031 01/2009

After adding the chlorine, let the water stand for about 30 minutes if the water is clear, or 1 hour if the water is very cold or murky.

DISINFECTING WATER WITH IODINE   Water purified with iodine may have a certain chemical taste to it; a vitamin C drink mix such as Tang will help neutralize the taste. Iodine-treated water may not be suitable for certain individuals—pregnant women, anyone with thyroid disease, or persons who are allergic to shellfish.

You will need:
❑ Water to be purified
❑ Clean container

❑ 2-percent tincture of iodine
❑ Medicine dropper

Place the water in a clean container. Add five drops of iodine per quart if the water is clear. If the water is cloudy, add 10 drops per quart.

To be absolutely sure the water is safe to drink, you may want to combine a couple of methods described above: filter out dirt, then treat with chlorine or iodine (not both), then boil and let cool.

25

LEARN TO DISTILL WATER

Distillation works by boiling water until it turns into water vapor or steam, then collecting it as it cools down and turns back into water. Distillation removes bacteria, salt, metals, and other chemicals in the water.

Distilling water yourself is fairly simple. You will need a large pot with a lid, a small cup, and a rope or string to tie a cup to the handle of the pot. Fill the pot about halfway with water. Turn the lid upside-down, and tie the cup to the lid's handle. Place the upside-down lid on the pot, making sure that the cup is not touching the water. Place the pot on your kitchen stove, a propane camping stove or whatever heat source you have available.

Distilling water at home

Boil the water for about 20 minutes. Turn off the heat and let it stand until cool. Don't lift the lid yet! The idea is to collect the water that became steam and turned back to water in the cup; opening the lid while it's still hot will interrupt the process. Wait until the pot is completely cool before carefully lifting the lid. You will find water in the cup that is safe to drink.

Another method you can use to distill water is to boil the water in a large pot and cover it with a clean thick absorbent cloth. The steam will rise and soak the cloth cover. As the cloth gets saturated, wring it out over a bucket or pitcher, and the result is drinkable distilled water.

26

LEARN TO BUILD A SOLAR STILL

If you are outdoors, with no pots or stove available, here's how you can distill water:

You will need:

- ❑ Shovel
- ❑ Bowl or wide-mouth container (this will be your collecting bowl)
- ❑ Leaves or vegetation (nonpoisonous)
- ❑ Plastic tarp
- ❑ Large rocks
- ❑ Small rock or weight

Dig a hole approximately two feet deep and three feet wide. Within the larger hole, dig a small hole and put the bowl or container in it. Fill the larger hole with leaves or other vegetation. Pour water, salty or fresh, along the side of the larger hole, making sure not to get any impure water into your bowl. Cover the larger hole with a plastic tarp. Secure the tarp along the sides of the hole with large rocks or other heavy objects. Place the small rock or weight in

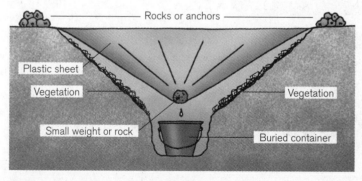

A solar still

the middle of the plastic tarp, directly over the collecting bowl. Leave your solar still in the sun for a few hours. As the sun heats up the vegetation, the water from the leaves and the impure water you pour will first evaporate and then condense on the plastic tarp. The weight of the rock will allow the condensation to flow downward and drip into your collection bowl.

27

LEARN TO BUILD A VEGETATION STILL

In an emergency, you can make a vegetation still if you are familiar with the plants and trees around you. You must be sure the tree or bush is nonpoisonous, or the water from that plant will be poisonous as well.

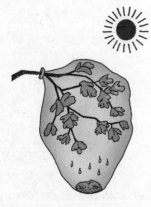

A vegetation still

All you need is a plant or tree in a sunny location, a clear plastic bag, a small stone, and a small piece or rope or band to secure the bag. Open the bag and let some air into it. Choose one side of the plant that receives the most sun and gather the leaves in that spot. Make sure there are no thorns or sharp branches that will puncture the bag. Place the bag on the gathered leaves until it is about three-fourths full. Place a small rock so the bag hangs down. Tie the bag shut, making sure it contains as much air as possible, with the stone end hanging lower than the mouth of the bag. This way, the moisture will drip down. After 24 hours, check the bag, loosen the tie, and collect the water. Choose another nonpoisonous plant the next day.

28

LEARN TO DISINFECT WATER WITH SUNLIGHT

If you are unable to boil water, you can purify it with sunlight. Fill clear plastic bottles with water that you have already filtered, and line a sunny area with heavy-duty aluminum foil. Lay the bottles down sideways on the aluminum so the shiny side of the aluminum foil can reflect sunlight. Make sure the bottles are facing toward the sun. Leave them out for six hours.

29

CHOOSE A COMMERCIAL WATER FILTER

Learning how to purify water yourself is a valuable skill, but at some point you may decide to buy a good commercial water filter for your home or a portable water filter to carry with you. Reverse osmosis is a common filtration method used in most residential or commercial water filters.

Here are some characteristics to look for in a water-filtration system:

- Easy to assemble and operate
- Easy to replace parts such as filters
- Reasonably priced to purchase and maintain
- Capable of removing bad taste and odors
- Can filter out dirt and sediment
- Removes bacteria, cysts, and parasites
- Removes metals
- Remove harmful chemicals, pesticides, and solvents

Consider how you plan to use the water filter. If you are staying put, you don't need to worry about weight and portability. But if you will be traveling or camping, you'll want a lightweight model.

Some good water filters include the Berkey Water Purifier, the Sawyer Inline Water Filter with 34-ounce bottle, the Katadyn Water Filter, and the Lifesaver Bottle.

30

CONSERVE WATER IN AN EMERGENCY

When water is scarce during an emergency, use every trick in the book to conserve it and make your stored water last as long as possible. It's also important to develop good water-conservation habits now, before a disaster strikes.

In an emergency, minimize water-drinking needs by staying as cool as possible and avoiding strenuous activity. Avoid eating salty foods so you don't get thirstier. Consume foods that have a high liquid content such as low-salt canned fruit and vegetables.

Use paper plates and cups and plastic utensils to avoid the need to wash dishes. Make sure you have a good supply of trash bags as well. Label each cup with the user's name and reuse the disposable cups throughout the day as long as they stay clean. Serve appropriate portions for drinking.

If you are unable to take showers, use a washcloth and a pan of water and take a sponge bath instead. Use anti-bacterial gel to clean your hands. Recycle your "gray water" (used water from laundry and bathing) for flushing the toilet, watering plants or cleaning floors. Reuse outerwear, sheets, and towels as long as possible. As long as they have not been contaminated, they are safe to use.

Teach your children not to waste water.

CHAPTER FOUR
Food Supplies

AFTER WATER STORAGE, FOOD SUPPLIES are also an integral part of your emergency-preparedness plan. If you've ever seen empty shelves at the supermarket after a "run" in anticipation of a storm or blizzard, and the store clerk tells you the items you want are out of stock, then you have witnessed a glitch in the "just in time" (JIT) inventory system. The JIT system used by most modern grocery stores relies on a sophisticated software program designed to calculate precisely when a supply will be depleted and have that item delivered just in time to replenish the empty shelf, thus eliminating the stores having a large stock of items in the supply room, avoiding spoilage and cutting down on inventory costs. This system works great on Super Bowl weekend and the day before Thanksgiving, but when an unforeseen event like a hurricane comes along, it just means you can't count on the store to carry extra inventory as a backup in the event the supply chain is disrupted. You have to do that yourself.

Consider what would happen if the transportation system gets disrupted due to a severe fuel shortage or a weather-related disaster. Stores would not get their deliveries even as shelves were emptied. If a large enough event were to occur, shutting down truck traffic on the interstates, there would not be any deliveries at all to stock the store. That's why they say we are all "six meals away from hunger"—because that is about as much as stores around us keep on hand.

31

TRACK WHAT YOU EAT

Before building your food storage supply, the first step is to track your food usage daily for 30 days. Every time you use a food item or ingredient, such as salt, sugar, or canned corn, write it down on a notepad. It doesn't matter how much you use, just write down everything. Include condiments and spices along with the main food ingredients. The list you come up with will tell you what foods you and your family eat on a regular basis, and this will be the list on which you will base your food storage plan. The next step is to take inventory of what you have in your refrigerator and pantry.

Food Usage List Example:

FREEZER

package of frozen corn

3 chicken breasts

4 steaks

1 frozen juice

REFRIGERATOR
10 eggs
2 sticks of butter
ketchup
4 yogurts
milk

PANTRY
2 cans peaches
2 chicken soups
2 cans chili
8 Pop-Tarts

Based on your list of the foods you know you actually use, calculate how many meals you have based on the number of people in your household. For example, if you have four people in the family, and you have the inventory above, you can have about three days worth of meals such as:

DAY 1
Breakfast: *4 Pop-Tarts, milk*
Lunch: *2 chicken soups, juice*
Dinner: *2 chicken breasts, corn*

DAY 2
Breakfast: *4 Pop-Tarts, milk*
Lunch: *2 cans of chili, juice*
Dinner: *4 steaks, corn*

DAY 3

Breakfast: *5 eggs*
Lunch: *yogurt*
Dinner: *scrambled eggs using the remaining 5 eggs and some of the butter*

In this simplified example, you can see that by the third day you are running out of food. If the stores are closed or you can't get there, and you have nothing else stored up, your family will start to grow hungry. This is what we are trying to avoid. You will need to build an emergency food supply in addition to what you have in your pantry. Knowing exactly what your family eats, you are already ahead of the game.

32

LEARN BASIC COOKING SKILLS

As a busy working mom, I'm constantly tempted to buy prepackaged foods and takeout meals for convenience. In many households, nobody cooks at all anymore, instead relying on frozen food or fast food for every meal. I have nothing against convenience foods, but constantly using prepared foods increases our reliance on the system and makes us vulnerable in the event of a disaster. If shipments of prepackaged foods were to grind to a halt, would we be able to cook foods on our own? The middle of an emergency is not a good time to start cooking lessons. Learn how to cook now while you have the luxury of time.

Basic cooking can be easy. It's only a matter of reading the recipe and following the directions. Endless recipes for every dish imaginable are available for free on the Internet. If you can follow directions on a box of macaroni and cheese or Hamburger Helper, you can learn to cook just about anything. The key is to start small and try easy recipes that do not require a lot of ingredients or special pans and utensils. Cooking also gives you control over the ingredients you use and lets you make healthier meals without all the extra salt, additives, and preservatives found in prepackaged or frozen foods. You don't even have to like to cook to learn how. If you only learn basic things like cooking rice or spaghetti, your repertoire will make you a lot less dependent on convenience foods, and you'll also save money.

33

START A FOOD STORAGE PLAN FOR $5 A WEEK

It can be intimidating to think about storing enough food for the family to last several months. Fortunately, you do not need to jump into a six-month food storage plan. Just start small and stockpile enough food to last for two weeks. Most families already have enough food for at least one week. Add to your stored food gradually.

When you do your weekly grocery shopping, in addition to your normal grocery list, pick up two to three items for your emergency stash. You can decide in advance how much money you want to spend for your food storage: $5 per week will get you three or four canned foods, a small bag of rice, and beans. To keep yourself from going over your budget, check the grocery fliers you get in the mail and the coupon inserts in the Sunday or Wednesday newspaper. Check the specials for the week and clip coupons for sale items that your family likes. Make sure you include a variety of foods that will keep well in storage such as:

CANNED MEATS
Chicken
Salmon
Tuna
Corned beef
Spam

OTHER PROTEINS
Beans
Other legumes like peas, peanuts, or lentils

CANNED FRUIT
Fruit cocktail
Mandarin oranges
Peaches
Pineapple

CANNED VEGETABLES

Corn

Green beans

STARCHES

Cereal

Crackers

Flour

Granola bars

Pasta

Ramen noodles

Rice

PREPARED MEALS

Canned pasta

Canned soups

Macaroni and cheese

Nonrefrigerated prepackaged meals

CONDIMENTS

Honey

Jelly or jam

Peanut butter

Pepper

Salt

Sugar

Syrup

Spices

For long-term food supplies, avoid frozen or refriger-ated items, which will spoil if the electricity goes out.

Here is an example of how you can start buying foods for your emergency supply without breaking your budget:

WEEK 1	ESTIMATED PRICE	TOTAL SPENT
4 cans tuna	$0.50 each	$2.00
2 boxes crackers	$1.00 each	$2.00
1 jar mayonnaise	$0.99	$0.99
1 jar pickle relish	$0.89	$0.89
WEEK 2		
4 cans chicken soup	$0.50 each	$2.00
1 box granola bars	$1.99	$1.99
WEEK 3		
2 cans corned beef hash	$0.99 each	$1.98
4 cans peaches	$0.75 each	$3.00
WEEK 4		
Flour	$1.99	$1.99
Sugar	$1.99	$1.99
Rice	$2.99	$2.99
Monthly total		*$21.82*

If you belong to a warehouse store, the next time you shop pick up a six-pack of your favorite canned soup, vege-table, or fruit. Pick up a different item the next time you go, and eventually you will have a nice stash of canned foods.

Round out your food inventory with spices and condiments such as soy sauce, ketchup, mustard, vinegar, and olive oil.

Keep buying a few extra items and in no time you will have a two-week supply of food. Continue until you have a month's worth of food and you will be well on your way. It is a good idea to keep up your food storage efforts until you have at least three to six months worth of food. It is entirely up to you, your available space, and your comfort level as to how much food you want to set aside.

34

BUY WHAT YOU EAT, EAT WHAT YOU STORE

Because you started the inventory list of what your family eats, you already know their likes and dislikes. When I was first starting my food storage program, I made the mistake of buying items with coupons or on sale, only to find that my family did not eat them.

During an emergency, kids may be apprehensive and nerves will be frayed. To help keep up morale, you'll want to include your family's favorite snack foods such as brownies, popcorn, and chocolate bars. Just be mindful of expiration dates.

35

STOCKPILE FOOD

Money-saving experts recommend stockpiling food as a great way to save on your household expenses. The trick is to buy a supply of the items you use while they are on sale and then not buy them again until the next sale.

Start collecting coupons and use them to purchase sale items. Combining coupons with sales will help you increase the savings, and you can buy more without making a dent in your budget. Take advantage of any rebates offered on items that you are already planning to buy. Add your rebate to your emergency savings or use it to boost your stockpile.

Don't wait until you are completely out of something to start shopping for it. Follow the preparedness rule: "Two is one, and one is none." If you have supplies on hand, you will never have to run to the store at an inconvenient time because you ran out of something.

Start looking at weekly store fliers and you will notice that supermarkets and discount stores typically offer their "loss leaders" (items priced very low to entice customers) on a schedule. Plan your shopping trips around these loss leaders.

Stockpiling may cost a bit more in the beginning while you are building your supply, but you will begin to see the savings after about a month.

Even if an emergency never happens, you will reap the benefits of having a stockpile with reduced grocery and fuel expenses once your plan is underway.

36

ROTATE YOUR FOOD SUPPLY

As you buy foods for your emergency pantry, take a marker and note the expiration date of the item on the top or front of the can so it is visible anywhere you store it. Expiration dates are printed on the container, but it is easier to mark them and make sure you don't end up passing the expiration dates. Pay attention to these dates as you review your inventory. Move older items to the front of the shelf to be used first.

I have been using this system for a while, and it is working well. I have not had to throw out any items due to expiration dates. If a date is getting close, I just use up the item or donate it to my local food bank.

If you run out of space in your main pantry, get a large plastic airtight container and start storing items in it. If budget is an issue, you can pick up airtight containers at garage sales. You can also use food-grade buckets that are often discarded by supermarket bakery departments. Keep containers in a cool dry place in your house, avoid storing food in the garage if you live in a hot, humid climate, as this will degrade your foodstuffs quickly.

37

KNOW THE DIFFERENT TYPES OF EXPIRATION DATES

Since you are planning on storing food for an extended period of time, you need to learn about expiration dates. They can be confusing and potentially misleading.

Slight differences in wording can change the meaning. For instance, "use by" means that the food should be consumed by this date; quality may decline after the date has passed. "Sell by" is more of an instruction for the store on how long they should keep selling the product. "Best if used by" or "Best before" indicates the quality of the food may decline after the date indicated but does not necessarily mean it will be unsafe or inedible after that date has passed, just that its flavor may not be the same.

Regarding canned foods, when shopping for your food stash, reach back deep into the shelf to find the cans with the longest shelf lives. Do not choose cans that are dented at the seams. They can explode in your car trunk or if stored in a warm place because the dent may put pressure on the seams, causing the contents to spill out. Avoid rusty cans, as rust can potentially allow bacteria to enter the can.

Throw out cans that are swollen, which can indicate botulism. Caused by toxins from a bacterium called Clostridium botulinum, botulism can be fatal.

Finally, make sure you have least two hand-crank can openers on hand. Canned food is useless if you can't open it.

38

CONSIDER STORING MREs

MREs, or Meals Ready to Eat, are used by the military and can be purchased online from several sources. Also known as field rations, MREs come complete with one of 24 eight-ounce entrées such as beef stew, chicken breast, pasta, or scrambled eggs, along with crackers and a spread such as cheese or peanut butter, a fruit-based beverage powder, and a dessert like cookies or candy, plus powdered coffee or tea, creamer, sugar, salt, matches, a plastic spoon, mint gum, and toilet paper, all packed in a plastic pouch. Kosher and vegetarian MREs are available. Each ration includes a disposable chemical heater. A case of a dozen MREs costs from $70 to $100. Stored in a cool, low-light place, they will last for five years. Because they were designed to be carried and eaten in the field, they are compact and lighter weight than canned foods. They are also dense in terms of calories and protein content.

39

STORE DEHYDRATED AND FREEZE-DRIED FOOD

Dehydrated and freeze-dried foods are great additions to your long-term food storage strategy. They have a very long shelf life. Some foods can last up to 20 years.

"Dehydrated" means water has been dried out of the food without cooking it completely. Methods used to dehydrate foods range from sun-drying and air-drying to baking in a kiln. Unless they come in a mix, dehydrated foods are usually bland and lack seasonings, so you'll want to keep your own herbs and spices on hand. At home, you can easily sun-dry your own foods such as fruits and vegetables or use a food dehydrator. Dehydrated fruit and vegetables can be stored in airtight containers, ideally with an oxygen absorber, or you can vacuum-seal them in plastic using a device such as a Seal-a-Meal or FoodSaver.

"Freeze-dried" means the food has gone through a flash-freezing process and was then processed with a miniscule amount of heat so the ice evaporated, removing all moisture from the product. Unlike dehydrated foods, freeze-dried foods are already seasoned, premixed with other ingredients, and precooked. This process allows food to stay fresh for a long time with very little loss in flavor; however, freeze-dried foods tend to be more expensive than dehydrated foods. You can't freeze-dry foods at home.

How to Use and Cook Dehydrated Food

Many types of dehydrated produce such as bananas, apples, and green beans can be eaten "as is." They are crunchy and flavorful. You can also add dehydrated fruit to cereal. Dehydrated strawberries and blueberries are delicious with hot or cold cereal and milk. Some dehydrated vegetables can also be eaten as a healthy snack with a bit of salt. To cook with dehydrated ingredients, first rehydrate and season the food.

You will need:
- ❑ Water—roughly 1 cup for each ½ cup food
- ❑ Container such as a pot, freezer bag, or thermos
- ❑ Salt and pepper, hot sauce, or other favorite seasonings

You can rehydrate dehydrated food simply by adding water, mixing, and leaving it for a few hours to soak. You will still need to cook it, as dehydrated food is still almost raw. This tends to take a long time, so a quicker way is to use boiling water.

THERMOS METHOD Boil water and add it to the dehydrated food in a thermos. Close the thermos, then leave it alone for a couple of hours. The food is essentially cooking in the thermos.

STOVETOP METHOD Boil water in a pot, add the dehydrated food, and cook until tender. Some foods such as dehydrated vegetables or potatoes will take only a few minutes, while other foods such as rice or beans may take longer. Either way, you will want to add seasonings such as salt and pepper or hot sauce to flavor the dish before eating.

40

DECIDE HOW MUCH FOOD TO STORE

Once you have a couple of weeks' supply of canned foods, try to increase it to a month's worth—and then, ideally, three months' worth. The actual amount you store is entirely up to you and your storage situation. If you feel ready to move to the next phase, calculate how much staple food such as rice, flour, sugar, and beans to store. Determine the quantities based on the number of people in your household and how much you expect will be consumed over a given length of time if circumstances prevent you from food shopping for an extended period.

For example, to calculate how much rice to store: A family of four—two adults and two teenagers—uses two cups of uncooked rice per meal. Multiply times two meals (if you expect to eat two rice meals a day). That's four cups of rice per day. Multiply times seven days—28 cups of rice per week—times four weeks to arrive at a total of 112 cups of rice, or about 45 pounds, in a month. That seems like a lot, but there are other variables to consider.

If you know you have other foods stored, you likely will not eat rice for two meals a day. To come up with a realistic plan, vary your meals and make sure your emergency stores include balanced quantities of the various food groups: grains, dairy, proteins, vegetables, and fruits. Consider space and budget when figuring how much of the various foods you can reasonably store for emergencies.

41

BUY STAPLES IN BULK

You can easily purchase and store your own bulk staples instead of buying them prepackaged from food storage suppliers. One advantage is that you will have control over the types of food and the packaging used for storage, so you know exactly what you have packed away.

Start by purchasing bulk food items such as rice, flour, pinto beans, and pasta. Many good deals can be found by visiting big-box membership stores such as Costco or Sam's Club. If you prefer not to join one of these stores, you can still hunt around for good deals at supermarkets, watching for grocery sale fliers and comparing prices between the various stores in your area. Another good source of bulk grains is ethnic markets. Depending on where you live, you may find large sacks of economy-priced dry staples at Mediterranean, Hispanic, or Asian markets.

Once you have your bulk items, you will need to purchase the materials for long-term storage:

❑ Five-gallon food-grade buckets with lids
❑ One-gallon-size Mylar bags
❑ Oxygen absorbers (300 cc)

The following common household items will also be needed:

❑ Iron
❑ Flat surface such as a wooden table with a cardboard liner
❑ Permanent markers for labeling
❑ Labels or masking tape so you can label the buckets
❑ Empty jar
❑ Measuring cup

Many experienced preppers recommend five-gallon-size Mylar bags because they fit into the buckets. I prefer one-gallon bags for staples because they are easier to transport than five-gallon bags. There is always a chance you may have to leave in a hurry. If so, all family members can carry one-gallon bags of food, while larger bags may be too heavy or unwieldy for smaller children. Because the goal is to keep the food fresh for as long as possible, if you open up a five-gallon bag, you'll need to use it all up. One-gallon bags can be opened and used as needed, without affecting the rest of the batches. If you wish, you can share one-gallon bags of food with neighbors or others in need without compromising the rest of the bin.

First, gather up all your food and labeling supplies:

- ❑ Food items such as rice, pasta, pinto beans, etc.
- ❑ Five-gallon food-grade buckets
- ❑ One-gallon-size Mylar bags
- ❑ Oxygen absorbers (300 cc)
- ❑ Measuring cup
- ❑ Iron
- ❑ Masking tape or self-sticking labels
- ❑ Permanent marker for labeling
- ❑ Cardboard to place over the table (under the iron)
- ❑ Airtight jar to keep extra oxygen absorbers
- ❑ Bay leaves to ward off weevils

Before you start, set the iron to the hottest setting. Make sure you place the iron upright on a covered surface to avoid burning. (I used recycled cardboard from a large pizza box.)

While these steps can be done by one person, it is easier to do them with two people so someone can hold the bag while the other person irons across it.

You will need to set aside a solid block of time to do this. Oxygen absorbers start activating as soon as you open the package, so if you have to stop and leave them for later, you must store them in an airtight container or they will become useless. *Do not* use oxygen absorbers for storing sugar, or the sugar will harden into a block and can only be softened again in a microwave, which is not always available.

Place one oxygen absorber in the bottom of the Mylar bag. Pour 12 cups of rice (or whatever you are storing) into the bag. Add another oxygen absorber and one bay leaf on top of the rice. There should be about an inch clearance from the rice to the top edge of the bag.

Line up the sides and across the top of the Mylar bag. Carefully iron across the top, leaving a one-inch space open on the left corner. Don't worry, the iron will not stick to the Mylar—it actually stays smooth. Do not try to make a fold across the top and iron it; it does not seal as well.

"Burp" the bag to let any remaining air out through the one-inch gap. Iron across the top 4 inches of the Mylar bag to seal it up. Careful—the iron can overheat. Some irons

Store bulk staple dry goods in Mylar bags

automatically shut off when they overheat. If yours happens to shut off while you are sealing bags, they will not seal as well, and it can require a good deal of waiting time before the iron comes back on.

Label the Mylar bag with a permanent marker and place it in the food-grade five-gallon bucket.

Keep packaging the same food item into Mylar bags following the above steps until the bucket is full. Then seal up the bucket and label it. Use a piece of masking tape or a self-sticking label and write the contents of the bucket with a permanent marker.

Store any remaining oxygen absorbers in an airtight jar. Store the food bucket in a cool, dry place—preferably not the garage. The next day, you will notice the bags look shrunken. This is the oxygen absorber doing its job.

42

OVERCOME THE ENEMIES OF FOOD STORAGE

To avoid waste, be aware of the potential problems related to food storage.

LIGHT Keep your food storage area dark. Light can degrade the taste and quality of stored food.

HEAT   Stored food should be kept below 72°F (22°C). Food kept at a higher temperature will begin to lose its nutritional value a lot quicker than if you kept it below 72. For every 18 degrees above 72°F the food will lose its nutritional value in half the time. The flavor and texture of the food will also be affected. Do not store canned goods close to heat sources.

HUMIDITY   Humidity will also negatively affect the taste, texture, and quality of the food. Keep all stored food off the floor, where humidity is most likely to seep in.

PESTS   Rice, flour, or pasta may get weevils. To keep bugs out of dry goods, store a bay leaf with these items inside your pantry. You may also freeze the bags overnight to prevent larvae from growing. Keep the area clean and free of pests. You also would not want any critters such as mice or cockroaches spoiling your food stash. Historically, cats were first domesticated for pest control, so if you have one as a family pet, let it earn its keep.

OXYGEN   Oxygen speeds up spoilage in food, so it is important to keep stored foods as airtight as possible. When storing foods yourself, make sure you use oxygen absorbers to preserve the life of your foods.

INACCESSIBILITY   Make sure your food storage is within easy reach and not in some faraway corner of your house or on a high shelf that is unreachable. You will be more likely to sort through and use foods that are close to their expiration dates if the food is easy to get to. Rotate your food supply constantly. Mentally plan on what foods you will use first in the event of an emergency. Be aware of what foods you have stored in the freezer, the refrigerator, and the pantry.

43

LEARN TO SPROUT SEEDS AND BEANS

Bean and alfalfa sprouts are usually eaten raw or minimally cooked in stir fries. They provide a nice crunch to dishes and are quite nutritious. It's easy to do your own sprouting. You will need to purchase seeds that are specifically designated for sprouting; they are not the same as seeds meant for planting. They should be organic, untreated with substances such as protective agents or pesticides, and not genetically modified. Many online vendors sell seeds for sprouting.

When you purchase your sprouting seeds, store them in a container labeled "for sprouting only." Do not seal them in an airtight container with an oxygen absorber. Seeds are alive, though dormant, and will die without air. Seeds that sit for too long will give off carbon dioxide that replaces the oxygen in the container. Open the container every few months to let fresh air in. Purchase fresh seeds every year.

Good sprouting seeds include mung beans, garbanzo beans, pinto beans, soybeans, alfalfa, broccoli, lentils, and radish seeds.

You will need:

- ❑ ½ cup of sprouting seeds such as mung beans
- ❑ 1-quart jar with a sprouting lid (These are just lids with holes in them to drain water easily.) You can also use a jar with a coffee filter on top held with a rubber band around it, or sprouting trays.
- ❑ ¾ cup water (Tap water is acceptable; you can use bottled water if you prefer.)

To sprout your seeds, measure about ½ cup of beans or other seeds and look through them to remove broken or rotten pieces or dirt particles. Pour the seeds into the jar and rinse. Drain out the water.

You are now ready to soak the seeds. Refill the jar with ¾ cup water and let it sit overnight—no more than 12 hours.

Drain out the water. It should not be smelly or foamy. Foamy water means the seeds have fermented, and you will need to start over.

When the water is drained, the seeds will be ready to grow. Leave the covered jar on its side in a sunny window or warm place. If you are using a sprouting tray, pour the seeds into the tray and cover them with paper towels or a dish towel.

Leave the jar alone for three to four days.

The sprouts are ready to eat when they are double the size of the seed. One-fourth cup of mung beans will provide about a cup of bean sprouts.

Eat them raw or stir-fry them as soon as possible. Sprouts will keep for less than a week in the refrigerator.

After that they will wilt. You can freeze them for a short time if you store them in airtight plastic bags before freezing.

44

LEARN TO COOK LENTILS AND BEANS

Beans and lentils are an excellent and inexpensive source of protein. They are a good addition to your food storage plan. Unfortunately, many people are not familiar with cooking beans or lentils so they avoid them. Don't miss out on a great food source—try these simple recipes.

Lentils

Lentils are so easy to cook because they do not need soaking and cook in about 30 minutes at the most.

You will need:
- ❑ 8 ounces (1½ cups) lentils
- ❑ 1¾ cups chicken, beef, or vegetable stock
- ❑ 1 bay leaf
- ❑ 1 onion, chopped
- ❑ 1 garlic clove, minced
- ❑ 1 small tomato, chopped
- ❑ salt and pepper, to taste
- ❑ 1 tablespoon oil

Place lentils on a plate or kitchen towel and remove any small stones or other debris. In a mesh colander, rinse the lentils in cool water.

In a medium pot over medium heat, warm the oil. Add the garlic, then add the onion and tomato and sauté for 3 minutes, or until the onion starts to soften. Add the lentils and the stock. Bring to a boil and cook for 20 to 30 minutes, until the lentils are tender, then season with salt and pepper. Serve with rice or bread.

Beans

Beans call for a similar cooking method to lentils but need to be soaked first and take longer to cook.

You will need:
- ❑ 8 ounces dry beans, such as pinto beans
- ❑ water
- ❑ 1 whole garlic clove
- ❑ ¼ onion (optional)
- ❑ 1 bay leaf

Sort through the beans and remove any stones or other debris. Soak the beans with one of the following methods; both work well.

OVERNIGHT SOAKING METHOD   In a large bowl, add enough water to the beans to cover them, then leave them to soak for at least 4 hours, or overnight. Drain the beans and rinse with fresh water.

QUICK SOAKING METHOD   Place the beans in an uncovered pot over high heat with enough water to cover them. When the water is boiling, lower the heat to medium, cover the pot, and allow to boil for 5 minutes, then turn off the heat. Leave the lid on the pot and let stand for 1 hour. Drain the beans and rinse with fresh water.

Once the beans have been soaked using either of the methods above, drain out the old liquid and add about 8 cups of fresh water to the beans. Add the garlic, onion, and bay leaf and bring to a boil over high heat. Lower the heat to medium and cook for 1 hour, or until the beans are tender. When the beans are cooked, add the salt (if you salt them too soon, they will be tough). You can also add your favorite seasonings such as ground cumin, chili powder, or Cajun spices. Serve with rice, tortillas, or bread.

45

LEARN TO MAKE HOMEMADE YOGURT

Although the easiest way to make your own yogurt is with a yogurt machine, you can make yogurt at home without one.

You will need (varies by method):
- ❏ Candy thermometer
- ❏ Large mixing bowl or casserole dish with a lid
- ❏ Wire whisk

- ❏ Glass or porcelain containers
- ❏ Bath towels
- ❏ Milk
- ❏ Plain yogurt
- ❏ Vanilla or other flavor
- ❏ Sweetener
- ❏ Heat source—You can use an oven, thermos, slow-cooker, or leave out in the sun.

OVEN Preheat oven to 100°F. Use the candy thermometer to check the temperature. Turn off the heat. Pour one quart of milk into a bowl or casserole dish. Add 3 tablespoons of plain yogurt. Stir well and cover the bowl or casserole. Let it sit overnight in the warm oven.

THERMOS Fill a thermos bottle close to the brim with warm milk that has been heated to 100°F (check the temperature with the candy thermometer). Add 2 tablespoons of plain yogurt. Mix well. Cover the thermos with the lid and wrap the thermos in two or three thick towels. Set the thermos in a warm dry place overnight.

OUT IN THE SUN Warm one quart of milk to 100°F. Pour the warm milk into a glass bowl or casserole. Add 3 tablespoons of plain yogurt and cover with a glass lid. Let it sit out in the sun on a warm (but not too hot) summer day. Let it sit for four to five hours. Make sure it is continuously under the sun and not in the shade.

SLOW-COOKER Turn the slow-cooker on low heat and pour in ½ gallon of milk. Heat the milk on low for 2½ hours. Turn off the slow-cooker and unplug, leaving the milk in the pot for 3 hours. After the 3 hours are up, remove 1 cup of the warm milk and place it in a separate bowl. Add ½ cup of plain yogurt and mix well. Pour this mixture back into the slow-cooker milk and use your wire whisk to mix well. Cover the slow-cooker and wrap it with two thick bath towels. Let the slowcooker sit overnight for 8 to 12 hours. The next morning, stir the yogurt and store it in glass containers. Refrigerate for eight hours before use. Flavor and sweeten the yogurt to your taste. You can also add fresh or frozen fruit.

46

LEARN TO MARINATE VEGETABLES

I used to think that food preservation would be too hard. Canning seemed intimidating, but it is actually a lot easier than it sounds. The easiest way to try your hand at preserving food is to marinate vegetables such as mushrooms or cucumbers. They are easy to start with and can be stored in the refrigerator for three to five days (though they're so tasty they may not last that long). Try these simple recipes.

MARINATED MUSHROOMS
2 tubs of mushrooms
2 garlic cloves, crushed
2 cups rice vinegar

½ teaspoon salt
½ cup sugar
Small bottle pimentos
¾ cup olive oil
Dash Tabasco sauce
½ onion, sliced

Clean and scrub the mushrooms to remove surface dirt. Mix the garlic, rice vinegar, salt, sugar, pimentos, olive oil, and Tabasco sauce in a bowl. Stir until the sugar is dissolved. Add the mushrooms and onions. Store in a glass jar and refrigerate overnight. They are now ready to eat.

"PICKLED" CUCUMBERS
Large cucumber
1 cup white vinegar
½ teaspoon salt
½ cup sugar
Pepper, to taste
½ cup vegetable oil
½ onion, sliced

Scrub the cucumber under the faucet. Slice into thin rings. Mix the vinegar, salt, sugar, pepper, and vegetable oil. Mix with the cucumber and onion. Store in a glass jar for about 4 hours. You can try this same recipe with other vegetables such as carrots, green beans, onions, peppers, or asparagus. Keep the jar in the refrigerator.

47

LEARN TO MAKE FREEZER JAM

When you have an abundant amount of fruit but are not yet
ready to delve into canning, make freezer jam. You don't
need a lot of special equipment, and you don't have to worry
about heating to the correct temperature. You do need to
make sure the jam is kept in the refrigerator or freezer and
used within the recommended time period.

STRAWBERRY FREEZER JAM
4 cups strawberries (3–4 pounds), cored and finely chopped
2 tablespoons lemon juice
1 cup superfine sugar
Fruit pectin
5 (8-ounce) wide-mouthed jars

Crush the strawberries with a fork in a large
bowl until juice is released. Add the lemon juice and
stir with a spoon until well mixed. Read the package
directions for the fruit pectin (they vary according to
brand) and prepare the sugar and pectin as directed.
When it is time to add the pectin and sugar mixture to
the strawberries, makes sure you stir constantly for 4
to 5 minutes to avoid lumps. Fill clean jars with jam to
about ¾ inch below the top so that the jam has room

to expand as it freezes. Jam will stay fresh for a week in the refrigerator or six months in the freezer.

48

LEARN BASIC CANNING METHODS

Canning is a way to preserve food by processing it at high temperatures for a long enough time that the heat kills any microorganisms that can cause the food to go bad. My grandmother used to can jams and jellies after the last harvest, and she continued to do it well into her nineties. The preserves were always such a treat. I thought canning was somewhat intimidating at first, but it is actually a simple process, which also involves vacuum sealing the jars to prevent air from entering the jar.

There are two basic types of canners: pressure canners and boiling water canners. Pressure canners are used for canning low acid-foods such as fresh vegetables (except

Pressure canner

Boiling water canner

tomatoes), red meats, poultry, and dairy. Low-acid foods need a higher temperature to preserve them than that produced by a boiling water canner. To kill the bacteria on low-acid foods, the temperature must reach and stay at 240°F, which the pressure canner does by creating steam under pressure.

Boiling water canners are more common than pressure canners, and they are used for preserving fruit for jams, jellies, and preserves, as well as tomatoes for spaghetti sauce or salsa. They reach a temperature of 212°F, sufficient to preserve these high-acid foods.

Let's talk about canning with a boiling water canner. You will need the following equipment:

BOILING WATER CANNER   A boiling water canner consists of a large pot, a rack, and a tight lid. It should be deep enough to immerse the jars, with an inch of water above the tops of the jars as it boils. Use the boiling water canner for fruits (jams, jellies, or marmalade) and tomatoes or other foods that have a high acid content. For other food items, you'll need to add lemon juice, vinegar, or another acidic liquid to the water.

CANNING JARS WITH LIDS   Canning jars and lids can be found in supermarkets, discount department stores, or hardware stores. Use standard canning jars and make sure they are not cracked or chipped. Use the canning lid that belongs with the jar according to the manufacturers' specifications. Lids are to be used only once; you can reuse the screw bands, but only if no rust or bending has occurred.

Other equipment:

- ❏ Jar lifter
- ❏ Rubber scraper
- ❏ Funnel
- ❏ Clean towels
- ❏ Wire rack
- ❏ Magnetic lid wand
- ❏ Ruler
- ❏ Miscellaneous cooking utensils such as cutting board, knives, colander, funnel, vegetable peeler, and paper towels.

Wash the canning jars with hot sudsy water, then rinse. Sterilize them by placing them in boiling water for at least 10 minutes. Continue to boil while you wait to fill the jars. Ready the lids and screw bands according to the manufacturer's instructions.

Add water to the canner and start heating it. Prepare the food you are planning to can. (See the sample recipes, page 97.) You will be canning the food while it is at boiling temperature.

Place the sterilized jars on clean towels before filling to keep them from falling or slipping. Carefully add the boiling preserves to the jars, leaving the recipe-specified headspace between the top of the food and top of jar. Use the rubber scraper to release any trapped air bubbles. Add more preserves to maintain the correct headspace.

Wipe the rims of the jars with a clean damp cloth. Any food particles left on the rim will keep you from sealing it properly. Place the lids and screw bands on the jars and tighten them according to manufacturer's directions.

Place the canner on the stovetop, half filled with water. Heat the water, then place the rack in the canner. Continue to heat the water so it is warm but not boiling. Place the filled jars in the canner, leaving space between jars so they do not touch each other. As you add jars, replace the canner cover each time. Once you put the last jar in, add more boiling water so that it covers at least one inch above the tops of the jars.

Cover and heat to boiling. Start your timer when the water is boiling. Keep the water at a gentle boil, and add more if the water level drops. Continue boiling according to the recipe.

Turn off the heat and remove the jars. Place them on wire racks to cool.

After about 12 hours, when the jars are cool, check the seal by pressing the center of each lid. If the lid moves up and down, the jar is not sealed. If the dip in the lid stays constant and you can't push it up and down, it is properly sealed. If some of the jars did not seal, refrigerate and use them within two days or freeze them.

Label and date the properly sealed jars and store them in a cool (no more than 70°F/21°C), dark, dry place. Use within one year.

Canning Safety

Before serving or consuming your canned goods, inspect the jars and lids thoroughly. Do not eat the contents if you observe that the lid is swollen, if you see mold in the food, if the food looks bubbly or murky, if the jar has leaked, or if the opened jar does not smell good. Discard the jar and food immediately if it does not look or smell good. Food that has been contaminated may contain harmful mold or bacteria that can cause botulism or other forms of food poisoning.

When you use the canned vegetables, boil them for at least 10 minutes before eating.

49

TRY SAMPLE CANNING RECIPES

STRAWBERRY JELLY

14 cups strawberries

1 package (about 2 ounces) powdered fruit pectin for low-sugar recipes

3¼ cups sugar, divided

1 tablespoon lemon juice

2 cups water

You will also need 6 (½-pint) canning jars and lids, the boiling water canner, a Dutch oven or large pot, cheesecloth, a colander for straining, and a metal spoon.

Wash berries and remove leaves. Use a masher to crush the berries.

Place the berries in the Dutch oven or large pot. Add ¾ cup water and bring it to a boil. Lower the heat, cover, and simmer for 5 minutes.

Layer a colander with three layers of cheesecloth. Place the strainer on top of another container that can catch the berry juice. Strain the berries through the cheesecloth and press to extract the juice. After you have extracted all the juice, add more water if necessary to total 4½ cups liquid. Place the juice back in the Dutch oven.

Mix the pectin and ¼ cup of sugar. Add pectin and lemon juice to the strawberry juice in the Dutch oven. Continue stirring as you bring to a boil, add 2¾ cups sugar and continue to stir. Bring to full boil for one minute, still stirring. Remove any foam that forms on top with the metal spoon.

Ladle the hot jelly into hot ½-pint canning jars, with ¼ inch headspace. Wipe the jar rims and seal. Boil the filled jars in the boiling water canner (as directed above) for 10 minutes. Remember, you start timing when the water starts to boil. Remove the jars from the canner, let cool, check the seals, and label.

TOMATO SALSA

20 medium tomatoes (about 10 pounds)
2 jalapeño peppers
10 poblano chile peppers
2 cups chopped onion
½ cup vinegar
½ cup chopped fresh cilantro
6 cloves garlic, minced
¾ tablespoon sugar
1 teaspoon salt
1 teaspoon pepper

You will also need 4 (1-pint) canning jars with lids, the boiling water canner, and a Dutch oven or large pot.

Boil the water in a pot. Dip the tomatoes in the boiling water for 1 minute, then rinse with cool water. Peel the skins from the tomatoes and remove the stems and seeds. Chop the tomatoes and drain them in a colander. You should have about 14 cups of chopped tomatoes. Place them in a Dutch oven and bring them to a boil. Reduce the heat and simmer for 45 minutes, stirring well, until the sauce thickens to a slightly chunky consistency.

Seed and chop the jalapeños and poblanos. Do not touch your eyes or mouth after you touch the chilies, as this can cause irritation. Add the peppers, onion, cilantro, vinegar, garlic, sugar, pepper, and salt to the tomatoes. Let the mixture boil, then remove from heat.

Pour the salsa into clean, hot canning jars, with ½ inch headspace. Wipe the jar rims and tighten the lids. Place the jars in a boiling water canner for 35 minutes. (Start the timer only *after* the water is boiling.)

Remove the jars from the canner with a jar lifter; let cool on wire racks. Label and date them.

50

DRY HERBS AND PEPPERS WITHOUT A FOOD DEHYDRATOR

If you don't have a food dehydrator, you can still dry herbs and peppers in a warm, dry, well-ventilated room. The temperature should be above 85°F, with low humidity (below 60 percent).

HERBS Pull up the herb plant from the soil. Remove any dead flowers or leaves. Rinse the plant gently in water to remove dust and dry thoroughly with a towel. Make sure no water is left, as it will cause mildew. Hang the bunch upside down in a roomy paper bag or exposed by itself; the leaves should be facing down. The area should be well ventilated, dry, and warm. When the herb feels brittle, pull the leaves off the stems over a piece of waxed paper or paper towel. Store in a cool, dry place.

CHILE PEPPERS You will need a large needle and string. You can use this method for red New Mexico chilies, Anaheim peppers, red jalapeños, poblano chilies, serrano chilies or xcatik Mayan chilies.

Wash and dry the peppers. Take the large needle and thread it with about a foot of string. Knot the end. Push the needle through the stems of the peppers, leaving at least ½ inch between each one. When the string is nearly full, tie a loop at the top. Hang the peppers in a well-ventilated spot in a warm, dark room. The pepper is fully dry when the skin feels brittle to the touch. Red jalapeños can also be dried in a smoker or barbecue over wood-chip coals to make chipotle chilies.

51

MAKE YOUR OWN BEEF JERKY

Beef jerky is a great source of protein and makes a good snack. Eat the jerky strips as is or use them as the meat ingredient for stews or soups. Soak them in water overnight to rehydrate before chopping them and adding them to the pot.

PEPPER JERKY

2 pounds beef top round or bottom round
3 cups beer
2 cups soy sauce
½ cup Worcestershire sauce
2 tablespoons black pepper

Wrap the beef in plastic wrap and allow to partially freeze for 2 hours. With a sharp knife, slice the beef with the grain and cut into ¼-inch slices.

In a bowl, mix the beer, soy sauce, Worcestershire sauce, and black pepper. Add the beef to the marinade, coating the pieces well. Cover and refrigerate for 8 hours.

Drain the marinade and pat the beef dry with paper towels.

Preheat the oven to 200°F. Rub the oven rack with oil. Place the strips on the rack, but keep the pieces from touching each other. Bake for 4 hours. The beef should be chewy but firm and dry. Let the beef pieces cool on the rack. Pack them in an airtight container.

This recipe makes ¾ pound of beef jerky.

SWEET AND SALTY JERKY

2 pounds beef eye round or bottom round
8 teaspoons sugar
8 teaspoons salt
2 teaspoons black pepper

Wrap the beef in plastic wrap and allow to partially freeze for 2 hours. With a sharp knife, slice the beef with the grain and cut into ¼-inch slices.

In a bowl, mix the sugar, salt, and black pepper. Lay the beef down on the wax paper on a flat surface. Sprinkle one half of the spice mixture on the beef. Turn the beef over and sprinkle the rest of the mixture. Mix well.

Preheat the oven to 200°F. Rub the oven rack with oil. Place the strips on the rack, but keep the pieces from touching each other. Bake for 4 hours. The beef should be chewy but firm and dry. Let the beef pieces cool on the rack. Pack them in an airtight container.

This recipe makes ¾ pound of beef jerky.

52

LEARN TO MAKE BREAD FROM SCRATCH

Learn to bake your own bread so you can control the ingredients and have hot, fresh bread anytime, even if bread supplies are low at the store following a disaster.

INGREDIENTS

1 cup water, warm but not hot (around 125°F)

½ tablespoon yeast

2 cups unbleached flour

1 teaspoon sugar
½ tablespoon salt
Extra-virgin olive oil

Mix the sugar and water. Add the yeast and let stand for 10 to 12 minutes, until the yeast blooms and the water is cloudy.

Mix the flour and salt together in a large bowl. After the 12 minutes are up, add the yeast mixture to the flour and mix well with a large wooden spoon. If the dough is too dry, add a bit more water. Add the 2 cups of flour gradually and keep mixing. The dough is ready when it is no longer sticky.

Leave the dough in the bowl. Brush the top with a light coating of extra-virgin olive oil. Set it in a warm place and let it rise until it doubles in size. Then shape the dough into two loaves and allow to rise a second time, until it doubles in size.

Preheat the oven to 425°F. Make a few slashes on the top of the loaves with a sharp knife before placing them in the oven.

Bake until the bread is done. Baking time can vary from 20 to 45 minutes, depending on the loaf size.

As you become proficient at making dough, save a cup of the dough and store in your refrigerator as "starter dough" for your next loaves. Tear the starter dough into small pieces and add to the new batch. This will give the bread a more complex flavor, like sourdough bread.

53

START AN INEXPENSIVE CONTAINER GARDEN

Growing your own food can be fun and easy. You don't need a lot of space to get started. You can start a garden anywhere that gets a few hours of sunlight, such as an apartment balcony. If you are new to gardening, start a small herb garden with parsley, cilantro, basil, and mint. Get a few supplies to start with and use recycled items like the following to save money:

EGGSHELLS AND EGG CARTONS   A week or two before you start your garden, begin collecting egg cartons and eggshells. When you cook with eggs, crack them along the middle of the shell so you are left with two half shells. Rinse the eggshells well and leave them out to dry.

HERB AND VEGETABLE SEEDS   Seeds are fairly inexpensive. Buy organic seeds from discount and home-improvement stores. You can also order heritage seeds or other non–genetically modified seeds online or from other gardeners who sell them at seed exchanges. Watch for local seed swaps where gardeners and small farmers trade different types of seeds.

GARDEN SOIL   Purchase garden soil from discount or home improvement stores.

POTS   To avoid spending a lot on pots, use five-gallon buckets with a drainage hole drilled in the bottoms. Many restaurants or bakeries give them out for free. You can also use empty coffee cans with holes punched in the bottoms for drainage, or purchase garden pots at garage sales and thrift stores.

DRAINAGE   Use rocks or pebbles or recycle Styrofoam peanuts to line the bottoms of your pots.

WATERING CAN   You do not need a store-bought watering can. Use an old water jug or clean milk jug.

WATER   Save money on water and avoid waste by collecting "gray" water from the sink, especially water you've used to rinse fruits and vegetables. Sudsy water from washing dishes is not recommended, as soap can be harsh on plants.

54

START YOUR OWN SEEDS

The first thing to do is find out your area's growing season. Prepare two months in advance. You can start sprouting seeds indoors in a well-lit spot. Take your dried eggshells and fill them with garden soil. Moisten the soil with water. Plant two or three seeds in each shell half. Place the filled eggshells in the egg carton and leave it in a well-lit, warm area. Keep the soil moist but not soggy; only water every two or three days. Once the seeds have sprouted about an inch high, they are ready to transplant into pots.

Fill the pot with garden soil to about two inches from the top. Dig a small hole and plant the sprouted seed, still in its eggshell, in the hole. Cover the roots with dirt, then water the soil under the new plant. Do not pour water directly on the seedling, as it can cause damage.

Water every other day or so, depending on your climate's humidity, so the soil is kept moist.

55

START A GARDEN IN A SMALL SPACE

One way to become more self-sufficient is to learn to grow your own food. Of course, unless you own a farm or have a huge backyard, there is no way you can meet all your food needs, but you can start a garden using various-sized containers in just about any space you happen to have: a small garden plot, an apartment balcony, or a small patio. Many cities and towns now have community garden projects where participants can grow their vegetables and herbs on shared public land.

Here are three considerations when starting your garden; once you have covered these gardening essentials, you're on your way:

SOIL   You need good soil to grow plants. If you have a small piece of land that already has rich fertile soil, you are in luck. If not, supplement the soil you have with good garden soil from your local home-improvement center. Choose the soil that is most appropriate for what you are trying to plant; some soils are best for container gardens, others are good for supplementing your natural soil or sand.

SUNLIGHT   The area you choose should have at least six hours of sun per day, as vegetables and fruit need lots of sun to thrive.

WATER   Plants need to be watered regularly, so unless you live in a rainy region, make sure you have a reliable water source. A garden plot can be equipped with automatic sprinklers or drip irrigation, or you can use a garden hose. If you have a balcony or container garden, you will need to water it on a regular basis.

Good tools to start with are a trowel, a small shovel, and a rake. Buying a few quality items is better in the long run than buying inexpensive ones that may break easily.

If you have never grown anything before, start small. Growing herbs in a pot is a good place to begin. A wide variety of herbs, including many that are not available in supermarkets, can usually be found in the spring at your local farmers' market. You can grow your favorite herbs on a sunny windowsill or right outside your kitchen and snip them right before using. You will enjoy the freshness.

When planting vegetables, choose those that your family normally eats. Tomatoes are a gardener's favorite, but consider others such as lettuce, peppers, and zucchini.

Dwarf fruit trees grow well in large pots or similar containers. If you are planting on a patio or balcony, make sure you check how much weight the patio or balcony can hold.

There is no reason you can't plant different varieties of plants next to each other—or flowers alongside vegetables. Companion planting is a way to save space and avoid pests. For example, marigolds keep pests away from tomato plants. You can plant shade-loving plants such as lettuce next to taller plants such as corn.

Check your local library for references about gardening. There is a world of information. An excellent resource about gardening in small spaces is Mel Bartholomew's *Square Foot Gardening* (Rodale). The Internet also has virtually infinite information for gardeners.

CHAPTER FIVE
Ready Your Home

WHEN FACED WITH MOST EMERGENCY situations, we all have a desire to stay close to our loved ones, protect our homes, and be surrounded by our own belongings. Unless a disaster forces us out, as was the case with the radiation leak in Fukushima, Japan, or Hurricane Katrina, most people would choose to stay home or at least plan to get home somehow. For this reason, we must take steps to ready our households for any potential disaster.

56

TEACH CHILDREN ABOUT PREPPING

When you start preparing on a regular basis, your young children will watch you gathering food for storage and items for emergency kits, and they will ask questions. Calmly explain that you are just making sure the family has enough food and water. Even if you are expecting an emergency such as a hurricane or ice storm, stay calm and relaxed when going about your activities so as not to alarm the children, who are often prone to exaggerate the danger in their minds. They will get used to the idea of preparing as it becomes a normal part of their lives.

If you are the parent of one or more infants and small children, add the following items to your emergency supplies:

- ❑ Diapers
- ❑ Wipes
- ❑ Sippy cup
- ❑ Pacifier
- ❑ Favorite blanket or stuffed animal
- ❑ Infant carrier
- ❑ Baby or children's clothes for layering: wool socks or booties, knit hat, gloves
- ❑ Thermal blanket
- ❑ Infant formula
- ❑ Baby food or dehydrated foods that cook into a soft consistency
- ❑ Baby soap and shampoo
- ❑ Antibacterial wipes or gel

Additional first aid items:

- ❑ Diaper-rash cream
- ❑ Lanolin if breast-feeding
- ❑ Baby Tylenol or Motrin
- ❑ Nasal bulb syringe

- ❏ Baby sunscreen
- ❏ Natural insect repellent that does not contain DEET
- ❏ Baby oil or Vaseline

57

DON'T FORGET ABOUT THE PETS

Pets are members of the family, so they must also be included in any emergency plans.

Include pet supplies in your emergency stash:
- ❏ Dog or cat food
- ❏ Extra water for pets
- ❏ Cat litter
- ❏ Flea and tick repellent
- ❏ Dog leash
- ❏ Food and water bowls
- ❏ Crate and bedding

Check with the Humane Society in your area to see if they have a pet readiness program. These programs offer safety tips for pets as well as free window signs for homes indicating to rescue workers that a pet may be inside.

If there is a foreseeable emergency that may require an evacuation, check with your county's Emergency Management Agency to find out which evacuation centers allow pets.

During extremely cold weather, keep your pets indoors even if they are normally used to being outside. If it's too cold out for humans, it's too cold for pets. If your dog must go outside for toilet needs, let him, but wait by the door so he can come in quickly. Be on the lookout for cats, dogs, or wildlife such as squirrels or rabbits sheltering under your car. If the engine has been running within the past few hours, it may be the warmest place around. Before starting the engine, bang on the hood or honk the horn to warn them, just in case.

Make sure your pet carrier is within easy reach so you can take your pet with you if you need to evacuate.

58

KEEP YOUR IMPORTANT DOCUMENTS IN ONE BINDER

To make sure all your important documents are in one place within easy reach, set aside a day to gather them all up. You will need only a sturdy binder and plastic protective sleeves.

Locate the following documents and assemble them in the binder:

- ❑ Birth certificates
- ❑ Passports
- ❑ Marriage license

- ❑ Copy of driver's license
- ❑ Personal records such as baptism and confirmation records
- ❑ Social Security cards
- ❑ School records—diplomas, report cards
- ❑ Vaccination records
- ❑ Vehicle ownership document or "pink slip"
- ❑ Car registration
- ❑ Credit card statements and other bills
- ❑ Printout of address book
- ❑ Insurance policies
- ❑ Checking and savings account statements
- ❑ Retirement account statements
- ❑ Mortgage documents or apartment lease
- ❑ Deeds
- ❑ Adoption documents
- ❑ Wills or trust documents

Personalize this list according to your own situation. Store the documents in a fireproof safe. You may also keep a backup set of copies (such as certified copies of birth certificates) in a separate location such as a safe-deposit box.

59

SECURE YOUR FAMILY PHOTOS AND KEEPSAKES

Most people would consider their family photos irreplaceable. To protect yourself from losing this valuable family history, make a project of scanning your photos onto photo CDs, DVDs, or memory sticks.

If your family is like mine, you have several sets of old photos taken before digital cameras came along. These photos cannot be replaced and are not backed up in the computer. We also have digital photos stored in several places: the hard drive, old e-mails, cell phones, and so on. Take a day or two to gather and organize these photos and scan them into data files. If you don't already have one, a computer scanner or all-in-one printer costs as little as $75 at any office supply store. A benefit to scanning your photos is you can preserve them before they fade or tear. Digital photos from cameras and cell phones can easily be backed up by saving the files onto CDs or DVDs periodically. I organize our photos by the year they were taken. If you keep scrapbooks, you can scan the pages onto a disk or memory stick as well.

60

MAKE A SAFE FROM A HOLLOWED-OUT BOOK

You will need:
- ❏ Hardbound book with at least 300 pages
- ❏ Ruler
- ❏ Pencil
- ❏ White glue
- ❏ Small bowl where you will mix water and glue
- ❏ Paintbrush
- ❏ X-Acto knife
- ❏ Rubber bands
- ❏ Page separators such as Post-its

Find a work area away from the reach of children. You will be using a sharp X-Acto knife to cut pages, so take safety precautions.

Select a hardbound book with at least 300 pages—one that most people will consider boring and leave untouched on a bookshelf. Turn about 75 pages and place a rubber band around them to separate these top pages from those you're going to glue and cut.

Mix 50 percent glue and 50 percent water in the small bowl. Stir well and add a bit more glue if needed so the mixture is not too runny. Brush the glue mixture along the outside of the book, but leave the rubber banded pages in front unglued. Let it dry. Separate the glued pages from the unglued pages by sticking the Post-it notes along the rubber band so they don't all get stuck together. Place a weight on top of the book while it dries.

Once the pages are dry, open the book at the first glued page. Using a ruler, measure a ½-inch margin from all edges of the book, including the fold, to form a rectangle. Then, with the ruler and X-Acto knife, carefully cut along the inside border all the way down to the back cover. Cut only a few pages at a time to keep the line as straight as possible. As you get to the end of the book, take care not to slice through the back cover. Discard all the waste paper and paper dust that remain in the book.

Brush the glue mixture on the inside of the book. Add a second coat of the glue mixture to the outside of the book. Keep the glued pages separate from the unglued pages. Let the book dry a second time.

A homemade book safe

Once the book is dry, close it and leave it alone for a day or so.

The secret book safe is now ready to hide your valuables. Remember to stack it in a bookcase with similar books that are not attractive to anyone.

61

ASSEMBLE A CAR EMERGENCY KIT

Your home should not be the only place you prepare for emergencies. Even if you are not going off on a long road trip, it's a good idea to make your own car emergency kit.

Keep these items in a container in the trunk:
- ❏ Three days' worth of nonperishable food like granola bars, MREs, and trail mix.
- ❏ Bottled water—four quarts or one gallon per person, per day for three days
- ❏ Flashlight and extra batteries
- ❏ Radio with batteries
- ❏ Matches
- ❏ Automotive tools, including jack and jumper cables
- ❏ Spare tire
- ❏ First aid kit
- ❏ Box of tissues or paper towels

- ❏ Emergency flares
- ❏ Umbrella or rain poncho
- ❏ Extra jackets
- ❏ Duct tape
- ❏ Emergency whistle
- ❏ Mini camp shovel
- ❏ Ice scraper and rock salt for snow
- ❏ Multi-tool
- ❏ Comfortable shoes, in case the car breaks down and you have to walk

Keep these items in the glove compartment:

- ❏ Maps and GPS device
- ❏ License and registration
- ❏ Emergency assistance card such as AAA
- ❏ Cell phone charger
- ❏ Emergency whistle
- ❏ Mace or pepper spray (keep away from the reach of children)
- ❏ Extra cash for emergencies
- ❏ Coins for toll roads

62

ASSEMBLE A DESK EMERGENCY KIT

An emergency can happen while you are at work, so it's a good idea to keep emergency supplies in a desk drawer or work locker. Besides tornadoes, earthquakes, and terrorist attacks, any number of everyday emergencies can occur, such as losing a button, getting a headache, or experiencing a power failure.

Include the following items in your desk emergency kit:
- ❏ Water (three to four bottles)
- ❏ Food: granola bars, oatmeal packets, crackers, trail mix, canned soup, ramen noodles, candy bars
- ❏ First aid kit: pain relievers, adhesive bandage, allergy medicine, antibiotic cream, antidiarrhea medicine, antiseptic spray
- ❏ Umbrella or rain poncho
- ❏ Emergency blanket
- ❏ Flashlight with extra batteries
- ❏ Compass
- ❏ Small sewing kit: safety pins, neutral-colored thread, a couple of needles
- ❏ Personal items such as dental floss, tampons, sanitary napkins

❑ Change of clothing

❑ Walking shoes, in case you have to walk home.

❑ A few dollar bills and change

63

ASSEMBLE A MINI SURVIVAL KIT IN A MINT TIN

For a mini survival kit you can easily carry around with you, gather the following items and store them in an empty mint container such as Altoids:

❑ Dental floss

❑ 1 tea bag and a sugar packet

❑ 2 small safety pins

❑ $20 in $1 bills and coins

❑ Band-Aids

❑ Aspirin, acetaminophen, or ibuprofen tablets

❑ Rubber band

❑ Small needle and thread

❑ Duct tape wound around a small pencil

❑ Alcohol prep pad or antibacterial wipe

Please note that this survival kit is just for minimum usage and is not intended to replace a first aid kit or complete survival kit.

64

STOCK UP ON MULTIPURPOSE ITEMS

DUCT TAPE This tape can temporarily repair a hole in just about anything—a tent, a shoe, a hose, or a taillight—anything that can use a temporary repair, duct tape can hold together. If the hem of your pants or skirt becomes undone, tape it up with duct tape until you can repair it. It can remove lint from your clothing or pet hair from a couch or seat cushion. You can use it to reseal packages, or you can trap flies or mosquitoes by hanging long strips of the tape on a window frame or rafters. People have reported success in using duct tape as a wart remover: Just cover the wart with duct tape, then replace as the tape falls off, and in a few days the wart should be gone.

WD-40 You can use WD-40 to remove duct tape and adhesives, remove bug remains from car surfaces and grills, and remove tar from cars and tire rims or scuff marks from floors. It will repel pigeons from perching on your window-sill—they do not like the smell of WD-40. Spray it on an umbrella stem to make it easy to open and close.

It will protect silver from tarnish, lubricate doors, windows and sliding door tracks, and silence squeaking door hinges.

BAKING SODA   Not just for baking, sodium bicarbonate, or baking soda, has multiple uses all over the house. Buy several boxes for both daily and emergency use. Keep a box next to the stove to extinguish minor grease and electrical fires. When baking soda heats up, it gives off carbon dioxide, which can smother flames by depriving them of oxygen. Stand back and sprinkle the soda liberally on the fire. You can use baking soda to gently clean baby equipment such as a high chair or stroller. Sprinkle directly on a moist sponge and wipe the surface, then rinse with water. Leave an open box of baking soda in the refrigerator or freezer to deodorize it for up to two months. When you replace the box in the fridge or freezer, pour the old baking soda down the drain and garbage disposal, following it with warm water, to get rid of odors and keep the drain free-flowing. Sprinkle it on a damp sponge and rinse with water to clean the refrigerator and microwave. Sprinkle it on cat litter or inside shoes to minimize odors. You can wash plastic containers with baking soda and water, use it to scrub pots and pans, or polish silverware with a paste of three parts baking soda to one part water, You can even use it as toothpaste.

For first aid, make a paste of baking soda and water and apply on the skin to treat insect bites, beestings, rashes, or poison ivy. To relieve diaper rash, place two tablespoons of baking soda in your baby's bath water. Relieve acid indigestion by adding one-half teaspoon of baking soda to one-half glass of water.

For cooking, add baking soda to water when cooking wild meat like rabbit or venison to remove the gamey taste. Add a pinch to iced tea to cut out the bitterness. Add a tiny pinch to scrambled eggs to make them extra fluffy. Use it as a leavening agent to allow dough or batter to rise. Scrub fruits and vegetables with it, then rinse with water, to remove germs or pesticide residue.

VINEGAR Clean windows by mixing two tablespoons of vinegar with a gallon of water. Fill a squirt bottle and spray it on glass and mirrors. Use old newspaper to wipe it off and prevent streaking. Clean the coffee machine by running a cycle with one-third vinegar and two-thirds water, then run another cycle with clean water. Clean the bathroom with full-strength vinegar to remove dirt, grime, and germs. Spraying vinegar on shower doors will also help remove hard water buildup. Do not spray on marble countertops, though, as the acid can harm the marble.

Clean the microwave by mixing one-half cup water and one-half cup vinegar in a microwaveable container. Microwave the liquid for one minute, then wipe down with paper towels. Minimize kitchen odors by leaving a small pan of vinegar on the counter. Clean greasy stovetops and barbecue grills with a mixture of one-half vinegar and one-half water in a spray bottle.

For a stain remover, mix one-third cup vinegar and two-thirds cup water in a spray bottle. Spray directly on the stain before washing. Spray it on full strength to remove underarm stains. Add one-fourth cup of vinegar to the final rinse to remove soap buildup on clothes.

Stop bug bites from itching by dabbing vinegar on the bite. White vinegar can also be used as an antiseptic for minor cuts and scrapes. Relieve sunburn by spraying with chilled white vinegar.

You can deter dogs from scratching their ears by wiping the area with undiluted vinegar, or keep cats away from certain areas by spraying the area with a one-half vinegar and one-half water mixture.

SALT   Clean your coffee pot with salt and ice cubes. It will remove coffee stains. If you drop a raw egg on a surface, pour salt on it, let stand for a few minutes then wipe it off. The salt makes the mess easier to pick up. When you have too much soap in the sink or washer, sprinkle with salt to reduce the excess suds. Use it as a scrub to clean greasy cookware. Freshen up a smelly cutting board by scrubbing it with salt, then rinsing with water. If a spill occurs inside your oven while baking, sprinkle salt on the spill and it will be easy to clean when the oven cools. If you spill red wine on a carpet, pouring a liberal amount of salt over it immediately will absorb it so you can simply vacuum up the stain when it dries.

For a natural weed killer, mix one part salt to three parts water in a spray bottle. Spray it on weeds growing in cracks on the sidewalk. Do not use on soil you will want to plant on, as it will also keep plants from growing. If you sprinkle a line of salt on a surface, ants will not cross it. If you pour salt on garden snails and slugs, they will melt.

To soothe a sore throat, gargle with a teaspoon of salt to a cup of warm water.

Relieve canker sores by rinsing your mouth with salt-water a few times a day.

Relieve beesting pain by moistening the sting with water and then sprinkling salt on the area.

When cooking, you can test an egg's freshness by placing one-half teaspoon of salt in a cup of water. Place the egg in the water. If it sinks, it is fresh; if it floats, it is old. Add a pinch of salt while making hard-boiled eggs, and the shells will come off easily. You can keep apples from browning by dipping them in saltwater.

WITCH HAZEL Witch hazel is an astringent made from the bark, leaves, and twigs of the witch hazel shrub. It was used by American Indians for medicinal purposes. Keep a bottle of witch hazel around for a variety of uses. It relieves rashes from poison ivy or poison oak. You can soothe painful varicose veins by placing a washcloth soaked in witch hazel directly on the veins. It also soothes external hemorrhoids. Dab it on acne or pimples to help dry them out. Place cotton balls moistened with witch hazel on under-eye circles to reduce puffiness. Soothe diaper rash by applying witch hazel with a cotton ball on the rash.

HYDROGEN PEROXIDE Commonly used to clean wounds and as a bleach, hydrogen peroxide is found in most households in 3 percent strength, diluted with water.

Keep a bottle or two in your supply cabinet. As a stain remover, it will remove fresh bloodstains or underarm stains from clothing. Apply a few drops of hydrogen peroxide directly on the stain and let it fizz. Then let it sit for a couple of hours before laundering as usual.

Sanitize countertops and surfaces in your kitchen and bathroom by mixing one-half cup water with one-half cup hydrogen peroxide in a spray bottle and spraying it on surfaces, then wiping with a dry towel. You can also use it as a fruit and vegetable wash to remove pesticides by mixing one-fourth cup hydrogen peroxide with two gallons of water in the sink. Rinse produce with plain water before consuming.

Whiten your whites with hydrogen peroxide as an alternative to bleach. Place in the bleach receptacle next time you wash white laundry. It can also be used to bleach hair blond.

Use hydrogen peroxide sparingly as a mouthwash to relieve blisters or canker sores, but do not swallow it. Taken internally, it can be poisonous.

Water your plants with one ounce of hydrogen peroxide mixed with a quart of water to help plants grow healthy roots and prevent root rot.

PETROLEUM JELLY   At $1 to $2 a tube or small jar, petroleum jelly is a bargain item with multiple uses. Besides moisturizing chapped hands and lips and preventing dry skin or windburn, it can help prevent diaper rash if you apply a thin coat on the skin. To keep ants away from pet food bowls, dab it on the rim of the food bowl. Ants will not cross a surface coated with petroleum jelly. Rub it along the rim of hard-to-open bottles such as nail polish to prevent sticking the next time you open them. Stop door hinges from squeaking by rubbing petroleum jelly on them. If your ring gets stuck to your finger, apply petroleum jelly to the area to help remove it. In a pinch, you can even use it as a makeup remover. If gum gets stuck in hair, use petroleum jelly to help remove it.

65

MAKE A CAN ORGANIZER FOR YOUR PANTRY

You will need:

- ❑ Empty cardboard fridge pack, like from a 12-pack of soda
- ❑ Scissors
- ❑ Glue
- ❑ Decorative paper or plain grocery bag paper

A homemade can organizer

Lay the cardboard fridge pack flat. The opening will be the front. Cut a rectangular hole on the top back end of the box. Cover the entire fridge pack with decorative paper of your choice. Starting from the back and working your way to the front, glue the paper onto the box and let dry. If you like, make a label for the front of the box. Load the cans from the back toward the front.

66

PREPARE FOR AN EARTHQUAKE

Because there are no advance warnings for earthquakes, if you live in an earthquake-prone area you must plan ahead and be aware of the hazards in and around your home.

If you have large objects such as paintings or picture frames hanging next to beds, move them away or reinforce them to make sure they do not fall on anyone. Secure any bookshelves or heavy objects such as refrigerators against the wall. Water heaters should also be secured with a support system against the wall. Repair any loose latches on cabinets or cupboards, as contents can spill out and fall on people during an earthquake. Store flammable liquids, pesticides, and poisonous substances securely inside latched cabinets.

Cut down any tree branches that are weak or dead to prevent them from falling and damaging your home or hurting someone. Fix any loose roof tiles or crumbling bricks on chimneys. Be aware of power lines around your property that could fall during an earthquake. Learn how to turn off your electricity, gas, and water utilities.

Keep sturdy shoes next to your bed.

Identify safe areas you can run to inside or outside the house—a sturdy dining table or desk or an open area away from tree branches, power lines or glass.

Create a plan on how you and your family can communicate and regroup after an earthquake. Choose a contact person outside your area whom family members can call.

Gather basic supplies:

- ❑ Portable food and water to cover all your family members for at least 72 hours
- ❑ Portable radio
- ❑ First aid kit and prescriptions
- ❑ First aid manual
- ❑ Flashlight
- ❑ Extra batteries
- ❑ Fire extinguisher
- ❑ Toilet paper
- ❑ Wrench you can use to turn off gas and water
- ❑ Hand-crank can opener
- ❑ Emergency contact numbers, including utilities, police, and fire department
- ❑ Portable toilet

DURING AN EARTHQUAKE   The general rule is "duck, cover, and hold." First, duck onto the floor. Take cover under a sturdy piece of furniture such as a desk or table, or in a strong doorframe. Then hold onto the furniture or doorframe and move or sway with it while the earth is still shaking. Do not move out from under it until the shaking stops. Stay calm and stay away from windows where glass can shatter, and from objects that can topple over such as mirrors and bookcases. After the shaking stops, leave the building, as aftershocks can still cause it to collapse. If you are outside during an earthquake, move to an open area away from power lines, trees, street signs, or streetlights. If you are driving, slow down and move to the side of the road in an open area. Avoid being under bridges, overpasses, power lines, or anything that can fall on your car.

AFTER AN EARTHQUAKE   Check for injuries and administer first aid. If you are barefoot, put on shoes with thick soles before walking anywhere. Look around for damage. If you smell or hear a gas leak, go outside, then report the leak to the gas company when you can. Do not light candles or use electrical appliances, as any flame or spark can cause a gas explosion. Check for downed power lines and stay away from them. Listen to the radio for news updates and warnings. If the power fails, unplug all electrical appliances to avoid damage from a power surge when it comes back on. Be prepared for aftershocks.

67

PREPARE FOR A HURRICANE

Be aware of hurricane watches or warnings: A *watch* means a hurricane is possible in your area, while a *warning* means the storm is expected to hit your area. Listen to radio and television announcements for new developments and possible evacuation orders. Learn about the five categories of hurricanes, based on wind speed, central pressure, and damage potential:

Saffir-Simpson Hurricane Scale			
Category	Sustained Winds (mph)	Damage	Storm Surge
1	74–95	**Minimal:** Unanchored mobile homes, vegetation and signs.	4–5 feet
2	96–110	**Moderate:** All mobile homes, roofs, small crafts destroyed; flooding.	6–8 feet
3	111–130	**Extensive:** Small buildings destroyed, low-lying roads cut off.	9–12 feet
4	131–155	**Extreme:** Roofs destroyed, trees down, roads cut off, mobile homes destroyed, beach homes flooded.	13–18 feet
5	More than 155	**Catastrophic:** Most buildings destroyed. Vegetation destroyed. Major roads cut off. Homes flooded.	Greater than 18 feet

From the Federal Emergency Management Agency website (www.ready.gov/america/beinformed/hurricanes.html):

- Decide where you will evacuate your family if necessary.
- Keep at least a minimum of a half tank of gasoline in your car in case you do have to evacuate.
- Learn how to turn off electricity, gas, and water utilities.
- If a hurricane is imminent, take the steps to prepare your home.
- Bring in all outdoor furniture, potted plants, garbage cans, or anything that is not secured to the ground.
- Trim tree branches that are dry or weak to avoid them falling on your roof.
- Board up your windows with plywood or hurricane shutters.
- Adjust your refrigerator to the coldest setting and keep the doors closed in case of power outage.

Gather your supplies in a bag so you are ready to evacuate if necessary. Include the following minimum items:

- ❏ Portable water and food—enough for 72 hours
- ❏ Portable radio and extra batteries
- ❏ Portable flashlight and extra batteries
- ❏ First aid kit including prescription medicines
- ❏ Maps
- ❏ Documents such as insurance policies and licenses
- ❏ Clothes
- ❏ Blankets
- ❏ Special needs as applicable such as pet food or infant formula

Prepare a communication plan so family members can get in touch with each other.

DURING A HURRICANE   Close all external doors securely. Close all interior doors. Stay inside a room in the interior part of your home, away from windows and exterior doors, preferably close to a restroom. Lie on the floor under a table or next to a heavy piece of furniture. Stay away from windows.

AFTER A HURRICANE   Monitor the news for the latest warnings and updates. Listen for any news of floods in your area. Return to your home only after the area is reported as safe. Survey the outdoor area for damage. Be alert for any loose power lines outside the home. Check for fallen trees, roof damage, or any instability in your home. Leave immediately if anything appears unsafe. Look for flooding or water damage. Take photos of all damage for any property insurance claims.

68

PREPARE FOR A TORNADO

Know your tornado terms: If there is a tornado *watch*, it means a tornado is possible in your area. A tornado *warning* means a tornado has been spotted by radar and you must find shelter immediately.

Be aware of tornado warning signs:
- Dark, possibly greenish sky
- Large hail
- Large, dark, low-lying cloud that may be rotating
- Loud roaring sound

DURING A TORNADO If you are at home, go to your basement with your tornado supplies. If you do not have a basement, head to an interior room or closet near the center of your house and stay there. Make sure you are dressed properly and wearing sturdy shoes. Do not open windows. There is a long-standing myth that all windows should be open during a tornado to avoid pressure buildup, but in fact this will result in flying debris or glass hitting you when the tornado arrives.

While the twister is hitting, get under a heavy table and cover up with blankets, sleeping bags, or anything that can protect you from flying objects. Protect your head and neck with your arms.

If you are in a car, trailer or mobile home, get out, find a building and head to the basement, or proceed to a designated tornado shelter. If you are outside and there is no shelter, do not try to outrun a tornado. Go into a low ditch, lie facedown, and protect the back of your head and neck with your arms. Stay as low as possible.

69

PREPARE FOR A FLOOD

If you live in a flood plain, purchase flood insurance. This is a separate coverage from your homeowners insurance. Losses from floods are not covered by homeowners insurance policies.

Make sure your furnace, water heater, and electric panel are elevated. Seal walls and basements with waterproofing materials to keep water from seeping in.

Pay attention to weather reports regarding flash floods, flood watches, advisories, or warnings. A flood *watch* means the weather experts expect a flood in the area, but it has not happened yet. A flood *warning* means flooding is going on. A flood *advisory* means the flood is either occurring or has already occurred and roads or streets may be underwater. In any of these cases, you must move to high ground and stay put.

If a flood is imminent, bring in all outdoor furniture and potted plants. If flooding is likely to occur, move small appliances to a higher floor and elevate larger ones if possible.

Gather your supplies in a bag so you are ready to evacuate if necessary. Include the following minimum items:
- ❏ Portable water and food—enough for 72 hours
- ❏ Portable radio and extra batteries
- ❏ Portable flashlight and extra batteries
- ❏ First aid kit including prescription medicines
- ❏ Maps
- ❏ Documents such as insurance policies and licenses
- ❏ Clothes
- ❏ Blankets
- ❏ Special needs as applicable, such as pet food or infant formula

If evacuation is imminent, turn off all electrical appliances. Turn off power at the main switch if it is safe to do so.

DURING A FLOOD   If you are driving, do not try to drive through flooded areas. A flooded street or intersection may look passable but turn out to be too deep for cars, which can become stranded or even wash away in the current. The only thing to do is turn back the way you came if you can, away from the flooded streets. In some cases, people cannot move either forward or backward and have no choice but to wait it out. Six inches of water is enough to stall most cars. A foot of water would cause many vehicles to float. If you are driving at night, be extra careful to avoid reported flooded areas, as it is harder to spot flooding in the dark. If you get stuck and floodwaters surround your car, leave the car if you can without wading through moving water and head to higher ground as safely and quickly as possible.

If you are on foot, do not try to wade through flood-water. A fast-moving flood can knock you down and carry you away.

If you are at home and you are standing in water or have gotten wet, do not touch any electrical appliances. Stay away from downed power lines.

AFTER A FLOOD   Return to your home after the flood-waters have receded and the area has been deemed safe by authorities. Be careful when re-entering a building that has been flooded, as the flood may have damaged the foundation.

Avoid coming in contact with floodwater, as it may contain toxic chemicals, bacteria, sewage, or gasoline, or may be electrically charged. If floodwater entered your home, clean and disinfect everything touched by floodwater. Report any downed power lines in your area and stay away from them. Listen to news reports regarding the safety of your water supply. Drink only bottled water until the water supply is declared safe to drink.

Have damaged sewer lines repaired as soon as possible.

70

PREPARE FOR AN ICE STORM

BEFORE AN ICE STORM Make sure you have a backup heat source such as a wood stove or gas or kerosene heater. Ensure that you have adequate firewood or fuel. If you have a generator, make sure you have plenty of fuel for it. Regularly check your generator and backup heat sources to make sure they are in good working order. When you have already lost power without other backups is not a good time to find out something doesn't work.

Make sure you have enough food and water. Review your 72-hour emergency kit and add enough supplies to last for a week. If you need to stock up on food, allow extra time for the shopping trip, bearing in mind that supermarkets can be extremely crowded when a big storm is approaching. Have adequate supplies of prescription medicines.

Buy extra batteries for flashlights and radios; keep extra candles and matches. Keep extra blankets. Fill up your cars' gas tanks and check the antifreeze in your vehicles. Charge up your cell phone. Check with elderly and special-needs family members or neighbors to see if they need anything.

Wash your laundry and dishes in advance. Have paper plates and utensils on hand. Have some board games, books, and other family activities available.

DURING AN ICE STORM   If the power goes off, turn off all appliances. You must have adequate ventilation when running a generator, wood stove, or gas heater.

Never run a generator inside the home; deadly fumes can result. Also, do not turn on gas stove burners to create heat; it can cause carbon monoxide poisoning.

Wear several layers of clothing for warmth. Drain water pipes to keep them from freezing, or open all faucets slightly to keep trickles of water running.

Stay indoors and avoid driving. Even if you consider yourself a skilled driver on snow and ice (and sometimes no amount of experience will help), expect that other drivers are not.

AFTER AN ICE STORM   Listen to weather forecasts and reports about road conditions. When you've concluded that it's safe to venture outside, clear areas around downspouts. When snow starts to melt, it will need to flow somewhere.

Assist elderly and other special-needs neighbors.

CHAPTER SIX

Personal Health and Safety

TAKING CARE OF PERSONAL HEALTH and safety is an important component of any preparedness plan.

Though there has not been a large-scale epidemic in years, there have been sporadic outbreaks of diseases such as swine flu or whooping cough in certain regions. Even a bout of seasonal flu that occurs every winter is enough to cause a small emergency in any household as germs get passed around and afflict everyone in the family. We also must be able to take care of ourselves and our families in case of minor injuries in the event of a disaster. The emergency room of a hospital would be overwhelmed with injured or sick people if there were a major disaster or pandemic. We should all take steps to improve our health and obtain our health supplies now while things are relatively calm.

We also all need to be aware that there is always a threat of crime, and it may get worse during a disaster. Being vigilant and aware may help you to avoid becoming a victim, but there are other habits that we need to adapt to overcome a false sense of security. A word about taking up defensive weapons: While being able to protect oneself and one's family is an important component of a preparedness plan, the choice of weapon is an extremely complex and personal decision. Because this is a basic beginner's preparedness guide, guns, knives, and other lethal defense weapons are beyond the scope of this book. Instead, the security section focuses on preventive steps we can take to avoid becoming crime victims.

71

FOLLOW THE NUMBER-ONE SAFETY RULE

The number-one rule of safety is very simple: Be aware of your surroundings. It's a habit that sounds so easy, but all too often we neglect it. Most people follow a daily routine, taking the same route to work and following the same schedule. Nothing bad ever happens, and it feels like nothing ever will, giving us a false sense of security. Insurance statistics show that most auto accidents occur within a few blocks from home. That's where most people feel safe and let their guard down. At the same time, thieves take advantage of your same sense of security. To increase awareness, actively look around when parking your car. Who is in the area, and what are they doing? Observe the area where you are walking and notice the people around you. Notice if anything or anyone looks out of place.

72

PROTECT YOURSELF FROM CRIME

Crime can happen at any time, but it tends to ramp up after a disaster. People can experience fear and vulnerability, and criminals may take advantage of the situation by looting and robbing stores, homes, or bystanders. To protect yourself from crime, develop a mind-set of avoiding danger. Practice safe habits at all times, even before any disaster happens. Being aware of your surroundings is a good start.

Trust your gut. If you get a sense that things "don't feel right," trust yourself and find a way to leave as soon as possible.

Think about what gets a thief's attention. To avoid being targeted, consider what might attract a criminal, such as flashy jewelry or a large purse that looks stuffed with cash or credit cards (even if it isn't). It does not matter what you have; you may just have coupons in your wallet, but if the thief *perceives* that you have lots of stuff, it's likely to get their attention. Look through your purse or wallet and consider whether each item is something you can leave behind.

Avoid having your belongings in plain view. When shopping, always lock your vehicle and do not leave your items in the car; lock them up in the trunk. When you're away from home, leave your curtains or blinds closed. The more items you have in plain view, the more likely a thief will take an interest in you.

Keep doors and windows locked. This may seem fundamental and self-evident, but a lot of people neglect to lock their homes or cars. Even if you are planning to be gone for "just a second," that is all it takes for a thief to enter.

Install a good alarm system. It will help protect you and your family from intruders.

Keep bushes and hedges around your home well trimmed. Overgrown bushes give a potential burglar extra "cover." They can crouch down and hide unseen behind tall shrubs and bushes. Keep all plants trimmed to a maximum height of three feet so you have a clear view of everything around you.

Look alert, ready, and confident. When walking to your car, have your keys ready in your hand so you don't have to fish around in your purse in the parking lot. Criminals can sense if someone will be an easy target, so don't look like one. Keys can also be an effective weapon of last resort for self-defense.

Train your children early. Teach them only to open the door to family or friends who know the "password" or special knock and never open the door to strangers.

Be selective about announcing your plans. Don't broadcast your activities or plans on social networking sites such as Twitter or Facebook. While you may trust your friends and family, you don't always know who else may be checking you out online.

Use common sense when you are out and about. Don't walk, jog, or bike when streets are deserted, such as early in the morning or late at night. If you do go out, go with a friend. Park only in well-lit areas close to your destination. Always lock your car, even if you are only getting out for a minute to buy gas or drop off mail.

Have a plan in case you do get followed or accosted. If you sense that someone may be following you, cross the street or change directions. Go toward a crowded place such as a restaurant or store and yell for help. If a stranger tries to pull you into their car, do everything you can to get away.

If you are thinking about buying a weapon, do your research and use common sense. Find out about laws in your state about gun ownership, concealed carrying, and lawful use. Get professional training on proper usage, safety, and storage of your gun, bearing in mind that firearms can kill or injure family members due to misuse, and they are among the most common items stolen in home burglaries.

Other options are stun guns and pepper spray, but both are restricted or prohibited in at least nine U.S. states, several major cities, and many foreign countries. You must research your choices carefully. And of course, keep in mind that you can never take any "nonlethal" weapon onto a commercial airliner. Regardless of what weapon you choose, keep all weapons away from children.

73

KNOW BASIC FIRE SAFETY

During a power outage, you may need to use candles or grills. Follow these basic safety tips:

CANDLES Keep candles away from curtains, drapes, and tablecloths. Blow out the candles when the adult leaves the room. Do not allow children or teens to light candles or burn incense unsupervised in their bedrooms. Keep all matches and fire starters out of the reach of children.

SPACE HEATERS Keep space heaters away from anything that can catch fire. Turn off space heaters before going to bed.

FIREPLACES AND CHIMNEYS Have your fireplace and chimney cleaned and inspected once a year.

COOKING SURFACES Never leave anything unattended on a lighted stove or barbecue. Stay in the kitchen or by the grill while cooking. Keep all dish towels, oven mitts, paper towels, and other potentially flammable items at least three feet away from the fire. Keep the barbecue grill away from shrubs or bushes.

FIRE EXTINGUISHERS Purchase a fire extinguisher for your home and learn how to use it.

SMOKE ALARMS Install smoke alarms in each room of your home. Change the batteries for your smoke alarms once a year. (If you're a smoker . . . quit! Then again, if you're a smoker, home safety may be the least of your problems.)

FIRE DRILLS Make a fire-escape plan with your family and practice it in a fire drill.

74

PROTECT YOUR FAMILY FROM GERMS

Epidemics come in all shapes and sizes, from common colds and flu to deadly outbreaks of hantavirus or plague. Exposure to disease-causing germs is usually unavoidable, but the chance of catching them can be minimized with simple precautions.

WASH YOUR HANDS OFTEN Make a habit of washing your hands as soon as you get home. This will ensure you are not bringing in germs from the outside. Always wash your hands before cooking, handling food, or eating. Teach kids how to properly wash their hands, too. Proper hand washing technique includes applying soap, rubbing the hands together, and washing them under the faucet long enough to sing a tune like "Happy Birthday." More germs are spread by handling money than by any other means, including public restrooms, so use hand santizier regularly when shopping. Avoid touching or scratching your eyes, nose, and mouth. Germs will enter your body through any of these openings.

KEEP YOUR KITCHEN CLEAN   Wash chopping boards and surfaces before and after cooking. Bacteria can get into food if it comes into contact with raw meat or other contaminants. A good way to disinfect surfaces is by using a mild bleach solution. When washing dishes by hand, keep in mind that dishrags and sponges can be breeding grounds for bacteria. Putting the rag or sponge in a microwave oven and heating it on high power for 30 seconds before using it will sterilize it.

DON'T TAKE ANTIBIOTICS UNNECESSARILY   Antibiotics cure bacterial infections, but they are ineffective against viral infections such as colds or flu. Your body builds up immunity to antibiotics the more often you take them. Take care of wounds and cuts as soon as possible. Clean the cut and apply an antiseptic like Bactine and an antibiotic ointment as soon as possible and protect it with a Band-Aid. If the cut lasts more than a couple of days without healing, worsens, or the area becomes red or swollen, or if a fever develops, see a doctor as soon as possible.

STAY HOME AND AVOID CROWDS   The best way to recover from illness is to get enough rest. Going out will only spread germs and infect others. If there is an epidemic or pandemic, stay away from public areas and crowds to avoid getting infected.

KEEP YOUR IMMUNE SYSTEM STRONG   Have you ever wondered why some people get sick a lot and others don't, even though they're exposed to the same germs? A strong immune system protects you from most diseases. Keep your immune system strong by eating healthy foods, getting enough sleep, exercising, and avoiding too much stress.

75

LEARN TO MAKE A DISINFECTANT CLEANER FROM EMERGENCY KIT ITEMS

You will need:
- ❏ Measuring cup
- ❏ ¼ cup bleach
- ❏ 2¼ cups water
- ❏ Gloves (optional)
- ❏ Clean 1-quart-size spray bottle or glass jar with a lid

In a well-ventilated room or outside, measure the bleach into your container. (Do not use a container that previously contained ammonia.) Avoid getting bleach on your hands; if you do, wash them under running water. Add the water. Place the lid on and gently roll the container back and forth to mix well.

Your cleaning solution is ready to use. This mixture is effective for disinfecting surfaces. If you have to make a larger batch, the Centers for Disease Control (CDC) recommends a 10:1 ratio of water to bleach (ten parts water to one part beach) Do not mix any other chemicals with this mixture. Discard after a day or two, as the mixture will no longer be effective. You can always make a new batch.

76

TAKE A FITNESS REALITY CHECK

How fit are you? If you had to run out of your house in a hurry and travel on foot with a small pack and kids, how far would you be able to get? Here are some tips to get you started on a fitness routine:

GET A PHYSICAL   If you haven't had one in a while, see your doctor for a checkup. Find out your baselines for weight, blood pressure, blood oxygen level, cholesterol, and other basic factors, as well as your body mass index (BMI).

BE KIND TO YOUR BODY   Start slow on your exercise routine. Walk or bike ten minutes a day initially, gradually increasing the pace and duration as you get used to it. Listen to your body. If you are getting too tired, or using weights is getting too uncomfortable, stop, go slower or use lighter weights. Don't forget to stretch before and after your workout. Drink plenty of water to stay hydrated.

STAY SAFE When walking or jogging, stay in known safe areas; let someone know where you are going and be aware of your surroundings.

77

ASSEMBLE AN INEXPENSIVE FIRST AID KIT

One of the most important steps you can take to ready your home and family for a disaster is to assemble a first aid kit. You can purchase a prepackaged first aid kit, but a more economical way is to assemble your own kit according to your family's needs. Another option would be to purchase the prepackaged kit as a base, then supplement with your own necessities.

You will need:
- ❑ Lunch box or small backpack
- ❑ Pain medication such as acetaminophen, aspirin (do not give to children), or ibuprophen
- ❑ Antidiarrhea medicine such as Imodium or its generic equivalent
- ❑ Antihistamine such as Benadryl, Zyrtec, or a generic equivalent
- ❑ Hydrocortisone cream
- ❑ Calamine lotion
- ❑ Aloe vera gel

- ❑ Charcoal tablets (use only if recommended by the poison control center)
- ❑ Syringe or teaspoon
- ❑ Bandages including wraps such as an Ace bandage
- ❑ Band-Aids in various sizes
- ❑ Antiseptic solution or wipes
- ❑ Burn cream
- ❑ Antibiotic ointment
- ❑ Cotton balls or cotton swabs
- ❑ Cold packs
- ❑ Gauze
- ❑ Scissors
- ❑ Tweezers
- ❑ Petroleum jelly
- ❑ Eyewash
- ❑ Eyedrops
- ❑ Extra eyeglasses, contact lens case, contact lens solution
- ❑ Disposable gloves
- ❑ Safety pins
- ❑ Hand sanitizer
- ❑ Thermometer
- ❑ Bulb syringe
- ❑ Duct tape
- ❑ Safety pins
- ❑ Needle and thread
- ❑ Moleskin padding (for foot blisters)
- ❑ Electrolyte solution (dry mix) (see page 156)

- ❏ First aid manual or reference book such as *Where There Is No Doctor* (Hesperian Foundation, revised edition, 1992)
- ❏ Your family's prescription medicines
- ❏ Flashlight with extra batteries

PRESCRIPTIONS   Try to keep at least an extra one- to three-month supply of your family's prescriptions such as asthma medicine and inhalers, blood-pressure medication, diabetes medication, and so on. You can do this by ordering an extra month in advance or by purchasing your prescription from a mail-order pharmacy. Let your doctor know you prefer to order your regular prescriptions in advance so he or she can write in an extra refill.

MONEY-SAVING IDEAS   As a container for your kit, you can use a child's lunch box or backpack. Or you can reuse something you have or purchase a container at a resale shop.

If you have friends or relatives who are also interested in building a first aid kit, split the cost; purchase the items in bulk and divide the supplies among yourselves. Fill small sandwich-size plastic bags with cotton balls, swabs, safety pins, and other small items.

Use coupons and take advantage of sales and special deals at pharmacies. Many pharmacies offer gift cards for new or transferred prescriptions. Bring the offer with you and use the free gift card to pay for your first aid supplies.

Many doctors' offices provide samples of over-the-counter and prescription meds. Don't be afraid to ask. There are also websites that offer free samples of over-the-counter products. Open an e-mail account just for this purpose and start signing up for freebies.

78

ASSEMBLE AN EMERGENCY DENTAL KIT

Most first aid kits leave out one important aspect: What do you do if you have a dental emergency just as a disaster is happening and you can't get to a dentist right away? A toothache or lost filling would be most inconvenient, not to mention painful. Put together an emergency dental kit with the following items:

- ❏ Cap or crown cement
- ❏ Temporary filling
- ❏ Dental floss
- ❏ Mouthwash
- ❏ Local oral anesthetic such as Orajel or Anbesol
- ❏ Denture repair kit if appropriate
- ❏ Cotton balls and Q-tips
- ❏ Clove oil
- ❏ Dental wax (if someone in your family wears braces)
- ❏ Toothpicks

A good book to have around is *Where There Is No Dentist* by Murray Dickson (Hesperian Foundation, 13th updated printing, 2010).

79

MAKE AN ELECTROLYTE SOLUTION TO PREVENT DEHYDRATION

You will need:

- ❏ 8 teaspoons sugar
- ❏ 1 teaspoon salt
- ❏ Flavored drink mix such as Kool-Aid
- ❏ Resealable plastic bags
- ❏ 5 cups water

Mix the sugar, salt, and flavored drink mix. (The drink mix is optional but will improve the taste greatly.) Assemble packets of the dry electrolyte mix in advance in sealed plastic bags and keep them in your first aid kit. When needed, add the mix to five cups of water. Mix well. Do not add more sugar than needed, as this may worsen dehydration. Drink the liquid slowly.

80

LEARN TO MAKE A GEL PACK FOR MUSCLE PAIN RELIEF

If you have sore or swollen muscles, here's how you can make your own gel packs to reduce swelling and pain.

You will need:

- ❑ Resealable plastic sandwich bag with a tight seal
- ❑ 1½ cups water
- ❑ ½ cup rubbing alcohol

Mix the water and alcohol inside the plastic bag and seal it shut. The bag should not be completely full. Freeze the bag for about four hours. The mixture will become slushy but will not freeze solid. Wrap a cloth or towel around it and apply to the swollen area. You can refreeze the pack after use.

81

LEARN TO MAKE A SALINE SPRAY FOR A STUFFY NOSE

You will need:
- ❏ Non-iodized table salt or sea salt
- ❏ Distilled water or plain tap water
- ❏ Sauce pan
- ❏ Small sterile bottle or bulb syringe
- ❏ Baking soda

Distilled water is recommended for this because it has no minerals or impurities. Distilled water does not have to be boiled. If you don't have distilled water, boil 1 cup tap water in a pan for 10 minutes. Let the water cool. Add ¼ teaspoon of non-iodized table salt or sea salt to the water. Add a pinch of baking soda to the mixture.

When the water is room temperature, pour it into a sterile nose-spray bottle or bulb syringe. Squirt three times into each nostril to moisturize and relieve dry or swollen nasal passages.

82

GET BY WITH WHAT YOU HAVE

TOILET PAPER   Ideally, we would all have a large stockpile of toilet paper, since this is one item every household uses in great quantities. In an emergency, though, you would have to ration this item to make it last as long as possible; and in a long-duration emergency, you might need to resort to alternatives when you run out. Here are a few possibilities:

Wet wipes or baby wipes are a great substitute for toilet paper, so plan on stockpiling when they are on sale.

Newspaper may work in a crisis, but the ink some newspapers use can turn everything black. Old phone directory pages or store catalogs may work better, since the ink used will not rub off. Just crumple the sheet until it softens up, then wipe.

Cloth—such as washcloths, terrycloth, old rags, or cloth diapers—is an environmentally friendly alternative to paper. Wet the cloth, wipe, and then launder the cloth. Since most people have nothing against rewashing cloth diapers, they should also be okay for bathroom hygiene. After use, throw the soiled washcloths into a bucket of water with some bleach to sanitize them and get rid of "skidmarks" before laundering.

Plant material such as sage leaves, cornhusks, or banana peel can be utilized as a toilet paper substitute. The plant must be nonpoisonous and nonirritating to the skin. The trick is to know in advance which plants are safe. You would not want to use something like poison ivy by mistake!

In many countries, use of the left hand in combination with pouring water with the right hand is the way to clean up. The idea is to clean vigorously using either a small container like an empty coffee can or a spray bottle, then dry with a towel. To avoid disease, thoroughly wash your hands with soap and water or with hand sanitizer right afterward.

SHAMPOO   Try to get by with less and less shampoo every time you wash your hair. Do not "lather, rinse, and repeat." Instead, if you are used to a quarter-size dollop of shampoo, try a nickel-size amount instead for a couple of days, then go down to a dime-size portion. Unless your hair is quite long, you will find it still comes out clean. You can also try washing your hair with water only and no shampoo. The hair will at least come out fresher than before. If you must wash with a soapy substitute, a tiny amount of dishwashing liquid will work. Just lather up the dishwashing liquid with some water and wash your hair with it. Rinse well.

To make a shampoo substitute, you will need:

- ❏ 2 empty shampoo bottles
- ❏ ½ cup baking soda
- ❏ 1¼ cups water, divided
- ❏ ¼ cup apple cider vinegar

In one clean, empty shampoo bottle, mix the baking soda and ½ cup water. Shake well to mix. In the other empty shampoo bottle, mix the apple cider vinegar and the remaining ¾ cup water. Shake well to mix.

To use, wet hair thoroughly. Rub hair and scalp with the baking soda and water mixture. Rinse with water. Then rinse with the apple cider vinegar and water mixture to prevent the baking soda from drying out the hair.

DEODORANT Natural crystal deodorant is available at drug stores and health food stores. It will last for a year with normal daily use. Just moisten with water and rub under the arms. Baby powder, baking soda, or cornstarch can also substitute for deodorant. Alcohol or antibacterial gel will work to kill bacteria that cause odor, but do not use after shaving or it will sting.

TOOTHPASTE AND MOUTHWASH Baking soda and water can be used as a toothpaste substitute. A teaspoon of salt mixed with a cup of water is a good mouthwash substitute. Follow with a clean water rinse.

SOAP A small amount of dishwashing detergent can be used as a soap substitute in a pinch.

To make your own soap substitute, you will need:

- ❑ Handful of oatmeal
- ❑ Water
- ❑ Washcloth

Mix a handful of oatmeal and some water in a washcloth. Rub on your skin and rinse well. This serves more as a scrub than as soap, but it will clean your skin without irritating it.

LAUNDRY DETERGENT Making your own laundry detergent will save you money. Or you can make your own detergent to make your store bought detergent last longer.

You will need:

- ❑ 1 bar bath soap such as Fels Naphtha, Ivory, or Zote
- ❑ 1 cup Arm and Hammer Super Washing Soda
- ❑ 1 cup borax
- ❑ Bowl
- ❑ Plastic container

Grate the bar of soap with a cheese grater and place it in the bowl. Add the borax and Arm and Hammer Super Washing Soda. Mix well. Store in the plastic container and label it. Keep it out of the reach of children.

To use, measure one tablespoon of homemade detergent per load, or use two tablespoons for heavily soiled loads.

83

LEARN TO MAKE A TEMPORARY TOILET

In an emergency, when water is unavailable or sewage lines are broken, do not try flushing your toilet, as it can back up and overflow into your bathroom. Make a temporary toilet instead.

You will need:

❏ 5-gallon bucket with lid
❏ Box of 100 large heavy-duty garbage bags
❏ Old toilet seat or two wooden boards
❏ Cat litter or other material such as dirt or sawdust
❏ Baking soda to help neutralize odors

Line the bucket with two layers of heavy-duty garbage bags. Place the toilet seat on top of the bucket, or place two wooden boards parallel to each other to make a seat. After use, pour in the cat litter or other absorbent material and some baking soda. Seal off the trash bag and replace it.

ALTERNATE METHOD   If you do not have a five-gallon bucket, you can drain the water out of your home toilet and line it with two heavy-duty garbage bags. Then follow the instructions above.

84

LEARN TO DISPOSE OF WASTE MATTER

In an extreme emergency, when water and sewage treatment plants are not available, you will need to dispose of human waste properly to avoid disease. Exposed waste will attract flies and other pests which will further spread disease.

You will need a shovel, gloves, and trash bags. Find a location that is at least 50 feet away from where people live or congregate and away from water. The area should be at enough of an elevation so water will not pool in it during a downpour. The idea is to prevent the contamination of water sources. Dig a hole at least three feet deep, or deep enough to cover the garbage bags completely. Bury the soiled garbage bags in the hole and cover them with soil.

CHAPTER SEVEN

When the Power Is Out

WE ALL TAKE ELECTRICITY FOR GRANTED, and the occasional power outage reminds us how much we rely on it. Power outages can be caused by any number of threats including natural disasters such as hurricanes, tornadoes, snowstorms, or earthquakes. Man-made problems can also cause power failures. Several years ago I lived in an area that had an outdated power grid, which had long been outgrown by the increased population in the community. We had frequent interruptions to our power service. Then there is the threat of terrorist attack on the power grid, an electromagnetic pulse (EMP) from a nuclear weapon, or the possibility of a solar flare strong enough to bring down the power grid. According to a science article on the NASA website, the sun can occasionally emit a "super solar flare" followed by an extreme geomagnetic storm, which can melt the copper wirings of transformers that support our electricity distribution systems. Power lines would act like antennae, spreading the current across a wide area. In 1989, a large solar storm disrupted power for six million people in Quebec.

85

ASSEMBLE A POWER FAILURE KIT

Designate a drawer or box and gather items in one place so you don't have to run around in the dark looking for flashlights when you have a blackout at night. Make sure the drawer or box is in a centrally located area that is easy to access in the dark.

You'll need these items in your power failure kit:
- ❑ Flashlights
- ❑ Light sticks
- ❑ Candles
- ❑ Matches
- ❑ Batteries of various sizes to fit in your flashlights and radios
- ❑ Battery charger and rechargeable batteries
- ❑ Radios, both battery-powered and hand-crank
- ❑ Warm blankets and clothes
- ❑ Propane or gas stove or other backup cooking methods
- ❑ Emergency food and water
- ❑ First aid kit
- ❑ Entertainment such as board games or a deck of cards

86

KNOW WHAT TO DO
IF YOU HAVE A POWER OUTAGE

SECURE YOUR FOOD SUPPLY Avoid opening the refrigerator and freezer doors unless absolutely necessary. Most food will stay frozen for about 24 hours in the freezer if the door stays closed. Cook or eat refrigerated items first, such as milk, eggs, yogurt, and other perishables. Once meats are thawed, they will need to be cooked and eaten right away. The easiest way to use refrigerated food is to make an omelet for breakfast, lunch, or dinner. Just chop up bits of meat or vegetables, sauté in butter, add to beaten eggs and cook on your camp stove.

STAY WARM Wear several layers of clothing to keep warm. Do not turn on a gas or propane stove to generate heat. Many people die each year from carbon monoxide poisoning as a result of turning on stoves to stay warm.

87

CONSIDER BUYING A GENERATOR

When we last had a power failure due to a hurricane, many households in our area were wishing they had a generator. It would certainly have come in handy. Home and garden stores generally have a shortage of generators when hurricane season approaches. Now is a good time to consider whether a generator is something you want to invest in.

There are several types of generators.

SOLAR Solar generators run on solar panels. They are more expensive and not as common as other types of generators, but they're safer and easier to maintain.

GASOLINE Generators that run on gasoline are the most common and easiest to find. However, gasoline can be tricky to store and transport due to the fire hazard it poses. Improper gasoline storage can also cause harmful fumes, and gasoline engines require more maintenance than other types of generators.

DIESEL Diesel engines are economical and low-maintenance but are noisy and smelly. Diesel fuel is also just as difficult to store and transport as gasoline, and it's not as readily available in some areas.

PROPANE OR NATURAL GAS   Propane and natural gas generators burn cleaner and longer than either gasoline or diesel.

It is best to hire a licensed electrician to install your generator to avoid any problems and accidents. Generators should be located outdoors in a shelter because of the noise and exhaust fumes. Never run a generator in an enclosed area, as there is a danger of carbon monoxide poisoning.

Follow the manufacturer's maintenance instructions for changing oil, fuel, or water. Before deciding to purchase a specific generator, make sure it is sold by a reputable manufacturer that has a service facility in your area.

88

LEARN TO STORE GASOLINE SAFELY

Gasoline is a hazardous substance, so you must be extremely careful if you decide to store it. First, be aware of the risks. Gasoline is highly toxic and contains known carcinogens such as benzene. Inhaling gasoline vapors is damaging to the lungs and central nervous system. It is highly flammable, so a small spark can ignite the vapors. Finally, spillage of gasoline on the ground will contaminate the soil and any nearby ground water.

To safely handle and store gasoline, use self-venting containers that are designated for fuel and have a safety vent, an opening designed to prevent the contents from reaching high temperatures and pressures that could cause an explosion. Store a small amount—less than ten gallons. Keep gasoline containers in a cool, dark place, away from direct sunlight. Make sure the storage area is well ventilated. Do not store gasoline containers in your basement or the trunk of your car. Keep containers away from the reach of children.

A proper gasoline storage container

Check for leaks at least once a month. Old gasoline becomes contaminated and unsafe to use after a while, so use it as quickly as possible by rotating your stored gas monthly. You can also use gasoline stabilizers such as Stabil; check with an automotive store before purchasing it. Dispose of old gas at a hazardous-waste collection site. Do not pour gas down the drain or on the ground.

Never smoke around gasoline. Avoid any skin contact with gas; wear nitrile gloves when handling it. Wash your hands thoroughly afterward. Never siphon gas with your mouth, as this can be fatal. Do not use gasoline for cleaning. Be careful to not overfill your containers or your gas tank.

89

LEARN TO COOK WITHOUT ELECTRICITY

There are several methods of cooking without electricity. It is a good idea to have several backup methods so that if one fails, or if you run out of fuel, you still have other means of cooking food.

GAS STOVE Many homes are equipped with natural gas and have gas stoves. If the power fails, most of the time the gas still works. However, in the aftermath of an earthquake, hurricane, or tornado, you should avoid turning on the gas stove because of possible leaks in either the gas main or your own gas lines. Make sure the area is well ventilated, and never turn on a gas stove to generate heat.

PORTABLE CAMP STOVE There are several types of portable camping stoves. The most common ones have two burners and run on propane canisters, which are usually sold in two- or four-packs. They are more versatile than the compact upright stoves made for backpacking. Buy several propane canisters as backups. Make sure you use the camp stove in a well-lit area.

ROCKET STOVE This type of wood-burning stove can cook efficiently using small pieces of wood. It is not as portable as a camp stove, but it's fuel-efficient and clean-burning.

SOLAR COOKER Various types of cooking devices use sunlight as their power source. Any food you normally cook on a stove or in an oven can be cooked in a solar cooker. The food does not need to be moved or stirred, and it cooks slowly, much like it would in a slow cooker. All you have to do is leave the food in the solar cooker outdoors for several hours, turning it occasionally so it constantly faces the sun.

A solar cooker

90

LEARN TO BUILD AN OUTDOOR PIT OVEN

If you want to make a makeshift oven outside, you can create a pit oven with easy-to-find materials:

- ❑ Shovel
- ❑ Large flat rocks the size of grapefruit
- ❑ Dry wood
- ❑ A large, moist canvas tarp

In firm ground, dig a square pit about three feet wide and three feet deep. Lay the rocks down in the pit in one layer. Set aside some of the rocks for use in sealing the tarp later.

Place the dry wood pieces on top of the rocks. Then build a fire on the wood and rock pile. Be careful not to get singed while leaning over the fire.

Make another layer of rocks to form a grate on which you can place the food you will be cooking. Lay the food on top of the rocks. You can cook meat on them, or put a pan of water on top of the rocks to boil food.

Cover the pit with the moist canvas tarp to make it heat faster. Secure the tarp by placing rocks on the edges.

Once you are done with the pit, throw sand and dirt on it or dampen the rocks to avoid fires. As with a campfire, never leave in unattended if a wind is blowing.

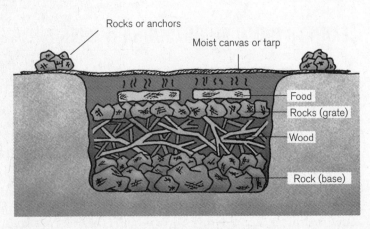

Rocks or anchors

Moist canvas or tarp

Food

Rocks (grate)

Wood

Rock (base)

A homemade outdoor pit oven

91

KNOW DIFFERENT WAYS TO START A FIRE

The ability to make a fire is an essential survival skill. Most experienced campers and survivalists become proficient at making fires; you can, too. In an emergency situation, a fire is necessary for cooking, warmth, and signaling your location.

You should stock up with several different means of starting a fire, such as:

- ❏ Matches, ideally the "strike anywhere" kind
- ❏ Magnesium flint striker
- ❏ Disposable lighters
- ❏ Butane lighter with plenty of fuel
- ❏ Tinder, kindling, and fuel wood or charcoal

To build a fire, first place a generous pile of tinder on the ground. Tinder is fine, wispy material such as dead grass, thin dry bark, small twigs, tissue paper, crumpled newspaper, or dryer lint that will light easily. Corn chips, because of the fat content, can also be used as tinder.

Next, cause a spark or flame that will ignite the tinder. Once the tinder has caught fire, blow on it gently so it can slowly grow into a larger fire. Then add kindling—larger twigs or wood shavings, and then dry branches from dead trees. Once the fire is burning, continue to add larger pieces of wood to fuel it.

To start a fire with steel wool and batteries you will need:
- ❏ 2 D batteries
- ❏ Steel wool
- ❏ Tinder, kindling, and fuel wood

Place the batteries end-to-end so that the positive (+) end of one battery is touching the negative (-) end of the other. Take a small amount of steel wool and stretch it out so that it touches the battery terminals at both ends. The steel wool will ignite quickly, so you will need to make sure your tinder is ready.

To start a fire with a lens you will need:
- ❏ Magnifying glass or eyeglass lens
- ❏ Sunlight
- ❏ Tinder, kindling, and fuel wood

Aim the lens so the heat of the sun is focused on the tinder. When the tinder ignites, start adding kindling, then fuel wood.

To start a fire with a reflector you will need:
- ❏ A flashlight
- ❏ Tinder, kindling, and fuel wood
- ❏ Sunlight

Take apart the flashlight and separate the reflector from the bulb, glass cover, and the body of the flashlight. Place the tinder in the reflector and set it in direct sunlight. The sun's rays should heat up the tinder enough to start a fire.

HOMEMADE FIRE STARTERS Smear petroleum jelly onto about 10 cotton balls. Save the petroleum jelly–soaked cotton balls in an old prescription bottle and label it *Fire Starter*.

Using a match or lighter, light a cotton ball and use it as tinder. The petroleum jelly will function as starter fuel.

92

CONSIDER ALTERNATIVE LIGHTING

Candles are a good source of lighting when there is no electricity. Make sure there is no risk of a gas leak, and keep candles far away from curtains, tablecloths, or anything else that can catch fire. It is also a good idea to keep some other light sources around:

FLASHLIGHTS Light-emitting diode (LED) flashlights are now quite common. They emit a bright light and are energy efficient. Keep hand-crank flashlights as a backup in case you run out of batteries.

TAP LIGHTS Tap lights come in several sizes and run on AAA or AA batteries. They are light and portable.

LED HEADLAMPS   LED headlamps for each member of the family, like the ones miners and cave explorers wear, are a great convenience. Since they are hands-free, they allow you to do chores like cooking in the dark.

SOLAR GARDEN LAMPS   Just leave them out in the sun all day and you have light at night. You don't even need batteries.

LIGHT STICKS   Stock up on light sticks when they go on sale after Halloween for a handy, cheap source of lighting in the event of short power interruptions.

CAMPING LANTERNS   Most camping lanterns are powered by C or D batteries, larger square lantern batteries, or propane canisters. The ones that use small batteries don't last long before it's time to replace them, so use them sparingly. Hand-crank lanterns are good to keep around for when you run out of batteries.

KEROSENE OR OIL LAMPS   These tend to be messy and require liquid fuels such as kerosene or lamp oil. They can be a fire hazard if you happen to spill the fuel, and if you let a kerosene lamp or heater run completely out of fuel, it will coat the walls of your house or tent with a layer of black soot.

93

LEARN TO MAKE A PERFECT CUP OF COFFEE WITHOUT ELECTRICITY

Assemble all your materials in advance.

You will need:
- ❏ "Green" (unroasted) coffee beans. The beans are actually tan colored when the reddish husk has been removed.
- ❏ Campfire popcorn popper or small covered skillet
- ❏ Manual coffee grinder
- ❏ French press
- ❏ Measuring cup
- ❏ Colander

ROASTING THE BEANS   Use a stove if you are trying this out with electricity, or try it off-grid on a propane camp stove. Do not use a camp stove indoors or without inadequate ventilation—it's unsafe and can cause a lethal carbon monoxide buildup. Roasting green coffee beans may cause a lot of smoke, which may set off your smoke alarm. You don't want a visit from the fire department while you are roasting your beans, or you might have to share the coffee with them! If you are roasting on a stove indoors, turn on the exhaust fan or open a window to make sure your area is well ventilated.

To start small, measure about a ¼ cup of green coffee beans. Preheat the popper or skillet over a low flame. Pour the green coffee beans into the skillet or popper, cover, and shake. Hold the popper steady for five seconds, then start jiggling it around. Keep the popper moving. When you start hearing a popping sound, lift the lid and look at the beans. The will start to turn brown after about five to seven minutes.

The popping is not constant like popcorn but happens every few seconds as the beans crack, giving off a bit of smoke. After about 10 minutes, check again and the beans will be brown. Turn off the fire.

You will notice some bits of chaff. Pour the roasted beans into a wire colander or just blow on them and the bits will fly off. Now you are ready to grind the beans.

Roasting fresh coffee beans

The green beans and the roasted beans are quite different, particularly the smell. In fact, the green beans do not smell like coffee at all. They have a pungent, plantlike smell, while the roasted ones have that strong familiar coffee smell. The aroma lingers long after you have finished roasting them.

GRINDING THE BEANS   Adjust the grinder for maximum coarseness if you will be using a French press. To do this, take off the handle, adjust the cog wheel up, and tighten it again. The grinder does not have any cushioning under the bottom, so you will need to stabilize it on the counter by placing a towel or pad underneath it. Remove the cork stopper from under the grinding mechanism. Pour in the beans and start grinding. Hold the grinder stable with the left hand and grind with your right hand (vice versa if you are left-handed). It will take some muscle power to grind the beans continuously and hold the grinder down. All in all, it should take about seven minutes to grind the ¼ cup of beans.

Manually grinding coffee beans

BREWING WITH THE FRENCH PRESS   The French press coffeemaker, developed in the 1980s by a small clarinet manufacturer in Normandy, is a simple way of making rich, full-bodied coffee and is now used in many fine dining

A French press

restaurants. I use the Bodum Shatterproof 8-cup French press coffeemaker, but any brand will do. Select a plastic or shatterproof model so it doesn't break accidentally. Eight cups sounds like a lot of coffee, but the "cup" actually means a European-style four-ounce demitasse cup, not an 8-ounce mug like most of us are used to. The ¼ cup of whole beans makes about two level 1½-tablespoon scoops of ground coffee. The rule of thumb is to use one scoop per four-ounce cup of coffee, though you can adjust the amount to taste.

Boil about 2½ cups of water in a separate pan. Turn off the flame once the water boils. If you use a plastic press, the instructions specify that the water must be hot but not boiling.

Remove the cover and plunger of the French press. Pour the ground coffee into the bottom of the press. Then add the hot water. Let it stand for four to five minutes.

Slowly insert the cover and plunger. Turn the lid so the pour spout is sealed and away from you. Press the plunger slowly until cannot go any further. Do not apply too much pressure, which can cause the water to splatter up or the coffeemaker to crack. If the plunger is hard to push down, it means the coffee grind is too fine.

Once the coffee is pressed, it is ready to drink.

These instructions will make a fresh, strong cup of coffee. At first it may seem like a lot of work, but it's worth it, and it will soon seem simpler than using a percolator or drip coffeemaker. With the price of coffee going up, it is cheaper to buy green coffee beans and roast them yourself than buying them roasted. Even if you never encounter an emergency where you need to make coffee off the grid, knowing how to roast and brew green coffee beans will help you save money while enjoying the flavor and aroma.

Of course, hardcore survivalists can make "cowboy coffee" by putting ground beans in a clean sock and boiling it in a saucepan, but quality is sacrificed.

CHAPTER EIGHT

When You Have to Get Out

MOST EMERGENCIES CAN BE WEATHERED by staying put in our homes and waiting until the crisis has passed. But there may be circumstances when you and your family will have to leave your home as quickly as possible. A few events in which the city authorities or emergency response teams may recommend or order evacuation include hurricanes, tsunamis, wildfires, potential flooding, and mudslides. You may also decide to leave if you feel your area is becoming unsafe or there is a widespread and extended breakdown of public utilities such as water or electricity. If you choose to evacuate, decide quickly, as the longer you delay, the more traffic congestion is likely to impede your efforts to get out, and if you wait too long, emergency conditions may prevent you from being able to leave at all.

94

PACK A BUG-OUT BAG

A bug-out bag includes all the supplies you might need when you have to leave your home. Each family's necessities are different, so tailor your list according to your own needs. When assembling the contents of your bug-out bag, you will need to consider the five survival essentials: food, water, fire, shelter, and security.

Once these basic needs are covered, you will also want additional items to make your stay more comfortable in a difficult situation.

BACKPACK Use a good-quality backpack, sturdy and as lightweight as possible. It should fit comfortably to your size and body type. A good backpack will have a hip belt so most of the weight sits on your hips and not your shoulders. It should have also have enough space and be reasonably priced. To avoid back strain, a person should carry a pack weighing no more than 30 percent of his or her weight.

CLOTHES Depending on the weather, you want to dress appropriately, in layers if possible. You may start out walking in the cool morning air but eventually get hot as the day wears on. You will want to remove layers as you go, then put them back on after sunset. Wool socks or synthetic fiber socks are better than cotton, as cotton retains moisture and will take a long time to dry if you sweat or get wet. Synthetic fibers like fleece are good, as they draw moisture away from the skin.

SHOES Hiking shoes are the best for walking long distances. You will need to break them in for a while before you have to bug out. It is a good idea to carry moleskin pads—self-sticking cotton flannel pads to place over sensitive areas before blisters can develop. You must take care of your feet, as they may potentially be your only mode of transportation.

MAP AND COMPASS Even if you have a GPS device, take a map and compass as backup, as there are certain areas where a GPS will not work well, such as valleys or parking garages. Bad weather can also interfere with them acquiring a satellite signal.

FIRST AID Personalize your first aid kit according to your needs as previously discussed in this book. I would want pain reliever, acid and diarrhea medicines, antibiotic cream for burns and insect bites, allergy medicine, wound care supplies, and tweezers, among other items. Bring nail clippers with you as well.

TENT You'll want the tent to be as lightweight as possible. A three-season tent will suffice, unless you are in an extremely cold area. If it is too bulky or heavy for one person, you can distribute pieces among your group to spread out the weight, as long as you stay together. You'll probably also want a ground sheet inside your tent to keep water away from you.

SLEEPING BAGS   Sleeping bags should be lightweight and versatile for various types of weather, and good to at least as low as 20°F. Down-fill bags are effective and comfortable, but they must not get wet. Synthetic-fill bags are not as comfy but will dry faster if they do get wet.

WATER AND WATER PURIFIER   You will need bottled water, as well as a way to purify more water. Take bleach and iodine tablets along.

FOOD   Take high-energy bars, ramen noodles, military-style MRE field rations, or dehydrated or freeze-dried foods. If you bring canned food, make sure you have a can opener.

COOKING AND EATING UTENSILS   The lighter the better, since you have to carry these things as well.

STOVE AND FIRE-STARTING MATERIALS   A means of cooking your food and boiling water is a necessity. Take a few lighters, matches, a camp stove, and fire-starting materials.

TOILET PAPER   Definitely bring a good supply of toilet paper for you and your family. You'll be glad you did.

WEAPONS   You might need to be able to protect yourself in a bug-out situation. A gun, crossbow, slingshot, or pepper spray are among the options, depending on your personal preferences. In many cases, the sharp knife you carry among your cooking utensils may be as good as anything else for self-defense.

DOCUMENTS   Bring all your family's important personal documents.

POCKETKNIFE AND OTHER TOOLS   Take a Swiss Army knife or multi-tool, a camp shovel, saw, an axe, hatchet, or machete, and perhaps a pick, as well as rope and paracord. On the road, basic car repair tools including jumper cables, adjustable wrenches, screwdrivers, and in some cases tire chains are also a must.

COMMUNICATION   Depending on the emergency situation, a hand-crank radio and your cell phone with a solar charger will help you stay connected. Take a pen and paper; a permanent marker may also be helpful for communicating with others.

OTHER USEFUL ITEMS   Sunglasses, sunscreen, and a wide-brimmed hat are all good for sun protection.

A multi-towel—a nifty little towel that takes the place of several bath towels and dries quickly—is not absolutely necessary, but it's nice to have if the budget allows.

Easier to use than a flashlight, a headlamp lets you keep your hands free in the dark to do other chores.

Bring insect repellent; commercial brands containing Deet are the most effective, but they are also corrosive and controversial—some studies show that they can cause seizures or neurological damage, and they should not be used on infants. There are also natural alternatives such as lemon-eucalyptus or citronella herbal formulas, which are not as effective or long-lasting as Deet.

Soap, toothbrushes, and other toiletries such as deodorant are not life-saving, but it's good to have them on hand for personal hygiene and morale.

Rain gear such as jackets and ponchos is important, as is a backpack cover, since most backpacks are not waterproof, and a rain fly for your tent.

Plastic bags and a deck of cards for entertainment will also come in handy.

This list does not cover absolutely everything needed to bug out, of course. It's a basic starter list you can add to. You may be thinking, "This could get expensive!" But you do not need to buy everything all at once. Just start picking up items little by little, and assemble the bug-out bag slowly as funds allow. Watch for close-out sales and garage sales, and search eBay and Craigslist for quality used items.

95

DETERMINE WHETHER YOU SHOULD STAY OR GO

A big question that comes up regarding preparedness is whether to shelter in place or leave in the event of a disaster. The answer may not be clear-cut, so it is important to consider several factors in making your decision.

TYPE OF DISASTER   The answer can differ according to what type of emergency you are facing. Some emergencies come with advance warning, such as hurricanes and forest fires. Even then the answer is not always obvious. A few years ago when the city of Houston was facing Hurricane Rita, many thousands of residents opted to evacuate all at the same time, resulting in gridlock and traffic jams, and the roads became impossible to get through. People were abandoning vehicles that had run out of gas or overheated, and many ended up getting stuck or turning back. Getting out became a mini-disaster all its own.

If you opt to get out, it's best to leave well in advance of the crowds. If there is an organized evacuation, you must be able to prepare your family and leave on a moment's notice. For many disasters, there is no advance warning, so you don't have a lot of choice as far as leaving ahead of time.

TRANSPORTATION What will be your mode of transportation? If you leave in a vehicle, are the roads passable? You need to make sure your vehicle is in good working order and that you have enough gas to get to your destination. Try to make a habit of keeping your gas tank at least at half full. Even if you are not preparing for an emergency, you will never be in a situation where you are desperate to find a gas station or risk being stranded on the side of the road. And if there ever were an emergency and you had to leave in a hurry, you could do so very quickly. Another benefit is that you can have bit of leeway as to where you fill your tank. You can find the cheaper gas stations before you actually need gas, instead of filling up at the first station you find no matter how expensive it is because your gas tank is empty.

SAFETY   Is it safe to be out on the road? If you have to leave on foot, would you be safe being out and about? For example, if there is a tornado approaching, that is not a time to be leaving your home.

YOUR FAMILY'S HEALTH AND ABILITY TO GET OUT   Are you in good enough shape that you can walk for miles with a load on your back? Do you have young children or elderly family members who may not be able to make a long trek to leave town?

INFRASTRUCTURE   You need to be aware of whether your town or city's infrastructure is still intact. If there are no utilities such as water or electricity, how long will it take to get them back? In the aftermath of Hurricane Ike, many Houston residents were forced to leave their homes and stay with relatives or at hotels until power and water were restored. If no infrastructure is left and there is no hope of restoring it any time soon, as was the case with Hurricane Katrina in New Orleans, you may have no choice but to evacuate—perhaps indefinitely.

These considerations will help you decide whether it is possible to get out or whether staying put is the best option. Hopefully you will never have to face this choice, but it is best of think ahead just in case.

96

MAKE AN EVACUATION PLAN IN ADVANCE

Preparing to evacuate can be stressful and panic-filled, but not if you have made a home-evacuation plan in advance. Assign responsibilities among family members to take the stress out of having to remember every detail yourself.

FOOD AND WATER   Take at least 72 hours' worth of food and water with you, more if your vehicle's space allows. You should have a variety of food: MREs, canned food (with a can opener), energy bars, dehydrated food, or food buckets with staples such as rice, beans, flour, and the like. Bring a few gallons of bottled water, along with your water-purifying system and water containers.

PERSONAL ITEMS   You should have your bug-out bag pre-packed and ready to go. Plan to bring a means of self-defense, emergency cash, and your first aid kit. If you have infants or very young children, bring enough supplies such as diapers, wipes, formula, special blankets or toys, and a pacifier. Bring any personal items you would not want to lose such as jewelry and family photos. Don't forget your prescription medications and any medical supplies such as a blood pressure monitor or diabetes strips, as applicable. Bring your hygiene supplies such as toilet paper and portable toilet. Take your personal computer, a charger—

preferably one that can work from your car's lighter socket while you drive—and your files backed up on one or more flash drives.

PET ITEMS   Pack your pets in their carriers and bring bowls, plenty of pet food, and extra water. Don't forget the dog leash and pet medication.

FAMILY DOCUMENTS   Bring your "grab and go" binder of documents.

READY YOUR HOME   If you decide to evacuate your home during an emergency, you will need to take care of a few things before you leave. In winter, these tasks may include draining water pipes and the water heater, pouring antifreeze on all water traps, and storing canned and bottled foods inside walls to prevent freezing.

At any time of year, you'll want to shut off the power. unplug electrical appliances, including computers, except for the refrigerator. Turn off the electricity at the circuit breaker box, except for the kitchen if you left the refrigerator on. Shut off the gas.

SECURITY   Lock all windows and doors, and put up storm windows, or board up or tape windows if necessary.

PREPARE YOUR VEHICLE   Get all your supplies packed in the car. Make sure your car is in good working condition and has enough fuel. Take extra fuel cans and oil with you if necessary. Have enough cash set aside for gasoline. Bring your spare tire and tool kit with you.

COMMUNICATION   Inform your immediate family of your emergency plans. Let your out-of-town relatives know your plans and destination.

Post the emergency plan in a visible area in your home, such as a bulletin board. Rehearse these steps with your family and see how much time it would take to get everyone out of the house. Aim for under an hour.

97

FIND EXIT ROUTES FROM THE CITY

Before an emergency happens, learn to read a map so you are not completely dependent on the GPS. GPS systems are a great invention but may not work in certain areas or if their map programs have not been updated recently. Figure out your location and find at least three alternate routes out of your area. Know how to get out by freeway, by main surface streets, and by side streets. Try the routes on a weekend so you are familiar with each drive. What if you have to leave on foot or by bicycle? Have a route in mind for each mode of transportation.

Acquire three types of maps:

- ❑ A local area map that covers your city and surrounding areas
- ❑ A state map
- ❑ Topographical maps showing elevation changes and including symbols for streams, vegetation, streets, and buildings

Maps can be obtained inexpensively or even free from many auto insurance companies and roadside assistance carriers. As of this writing, they can also be downloaded and printed from online resources such as Digital-Topo-Maps.com. Another free resource is Google Earth. Keep a set of maps in your car and in your bug-out bag at all times.

98

HAVE THE RIGHT FOOTWEAR FOR WALKING OUT

If there is an emergency situation and you have to walk out, through debris, flooded streets, or whatever crisis awaits, you definitely want the most comfortable, protective footwear possible. This means a closed-toe pair of shoes with durable soles. Comfy sandals or slip-ons won't fit the bill. Choose the shoes with the best fit. A good fit means they are roomy enough that you can wiggle your toes, but not so loose that your feet slide around inside. They should feel firm around your feet, but not too constricting.

Hiking shoes are highly recommended for long treks. They come in low-cut, mid-cut, and high-cut; high-cut provides the most ankle support. The shoes can be made of leather, synthetic material, or a combination. Leather shoes tend to be more durable but require more time to break in and are more expensive; synthetic materials are lighter and tend to cost a bit less, but may not last as long.

Whatever material or price, make sure the shoes are already broken in. Hiking out of the city in a pair of new shoes will cause blisters. Your survival could depend on your ability to walk for a long period of time, and shoes that do not fit or that your feet are not accustomed to could delay or prevent you from getting to safety. Here are a few tips on breaking in your shoes for maximum comfort:

Make sure you have good socks; wool is best. Wear the shoes around the house for a couple of days, then wear them when you go on short errands for another three days. If you feel a blister coming on, cover it with a Band-Aid or moleskin pad in the area of discomfort, hopefully before the blister forms. Do not wait until you are in pain.

As you get used to the shoes, go on a short walk around your neighborhood. Finally, take some small hikes, starting with a mile or two, then go on longer hikes when you start feeling comfortable.

Some sources believe you should break in shoes the quick way by submerging them in water then letting them dry around your feet. However, this may shorten the life of your shoes, so I do not recommend it.

During a crisis you won't have time to be searching for just the right pair of shoes, so do it now while nothing is going on, and leave the shoes in an easy-to-reach spot by the door.

99

KEEP PARACORD ON HAND

Paracord is actually parachute cord, also known as 550 cord because it is rated to hold 550 pounds. This lightweight, fairly elastic nylon cord was originally used in parachutes in World War II. It dries quickly and resists mildew. Paracord is composed of an interior and exterior: the inner strands are composed of seven two-ply nylon strands; on the exterior is a woven nylon sheath. It can be used as is or taken apart for a variety of uses.

Here are a few ways to use paracord:
- Make a shelter by tying branches together.
- Hang a food bag from a tree to keep it away from bears.
- Use it as fishing line or clothesline.
- Secure items to your car.

- Replace shoe or boot laces.
- Climb.
- Hoist a load.
- Use it as an animal leash.
- Secure planks to build a raft.

A good way to store paracord is to wrap it around the handle of an axe, knife, or other tool. It cushions the grip and can be unraveled when you need it for other uses. You can also wear it around your wrist and forearm like a bracelet. Paracord can be purchased in 100-foot lengths in a variety of colors, including olive drab, orange, black, or camouflage.

100

LEARN TO NAVIGATE WITHOUT A COMPASS

If you become stranded somewhere without a compass, you can tell directions using one of these techniques:

LOOK AT THE STARS If you are in an area where you can see stars, use them to pinpoint a general direction. In the Northern Hemisphere, find the Big Dipper constellation in the sky. The two stars that make up the front of the dipper point toward the North Star, the bright star roughly four times the distance between the two stars that line up on it. The North Star does not move in the sky as other constellations rotate around it, so it's always due north.

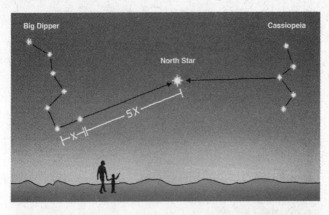

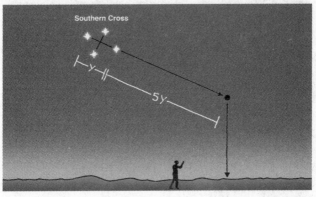

Using the North Star (top) and the Southern Cross (middle) to pinpoint direction. Bottom: Using an analog watch to determine direction in the Northern Hemisphere (left) and Southern Hemisphere (right)

In the Southern Hemisphere, find the Southern Cross constellation in the sky. Then look for the two stars below the cross. Imagine a line from the cross and a line extending from the midpoint between the two stars. Extend the lines to find where they intersect at a right angle. This points south.

ANALOG WRISTWATCH If you are wearing an analog watch and the sun is visible, you can find out which way north and south are. In the Northern Hemisphere, point the hour hand of the watch toward the direction of the sun. The midpoint between the hour hand and 12 o'clock is south. If you are in the Southern Hemisphere, point the watch's 12 o'clock mark toward the sun. The midpoint between 12 and the hour hand is north.

101

LEARN TO SIGNAL FOR HELP IF YOU ARE STRANDED

While not always feasible, it is a good idea to notify friends and family of the time when you are leaving and your destination. This way, if you do not arrive at the appointed time, someone will be aware that you may have gotten stuck somewhere.

If you are stuck in your vehicle by the side of the road and you do not have a cell phone to call for help, lift up the trunk of your car as well as the hood. This will indicate to potential rescuers that you need help. Another recognized signal that you need help is to tie a red piece of cloth to your car's antenna.

If you have a radio or other ability to signal with noise, any signal that comes in threes is universally recognized as a call for help. A signal that comes in twos is the rescuer's signal either responding to yours or announcing that help is in the area. If you become stranded in a remote area, you can indicate you need help by building three fires in a triangle pattern. The brighter the fire, the better you will be spotted at night. In the daytime, rescuers will more easily spot a smoking fire; piling green vegetation on it will make more smoke. Or, using sticks, rocks, branches, or any materials available, make large letters—at least 12 feet tall—on the ground in a clearing or field, spelling HELP or SOS.

The following symbols signify these messages according to Ground-to-Air Emergency Codes:

Code Symbol	Message
V	Require Assistance
X	Require Medical Assistance
---------------->	Proceeding in this Direction
Y	Yes
N	No
F	Need Food and Water
LL	All Is Well

OTHER POSSIBLE SIGNALS Blow on a whistle three times, one after another, with a few seconds interval in between each blast. Wait a few minutes, then repeat the sequence. Or, if the sun is out, use a mirror or other shiny object such as aluminum foil, a shiny belt buckle, or a soda can to reflect the sunlight. Flash the light toward one area in the universal SOS pattern—three short flashes, three long flashes, three short flashes. According to the US Army Survival Manual, "Do not direct the beam of light toward an airplane's cockpit for more than a few seconds, as this may blind the pilot. Do not make the flashes too rapidly as a pilot may mistakenly interpret these as enemy flashes."

At night, you can make the SOS signal by using a flashlight and directing the beam toward a passing aircraft. You will want to get to an area that is as high as possible for the best chance at being spotted.

You can also make signals with your own body. To signal that you need urgent medical assistance, lay on the ground with your arms above your head. To signal that you need help, while standing up, raise both arms with open palms.

A Final Word

I WROTE THIS BOOK FOR THOSE who have some interest in being prepared and knowing what initial steps to take. I hope you have found some ideas to inspire you to become more self-sufficient. Even if you try only a handful of tips, you will reap some benefits and peace of mind knowing your family is prepared.

Resources

Web Sources

apartmentprepper.com

des.nh.gov

geology.com/articles/earthquake-safety.shtml

thesurvivalmom.com

www.businessweek.com/bwdaily/dnflash/content/
oct2006/db20061002_959305.htm

www.consumeraffairs.com/news04/2009/08/expiration_
dates.html

www.popularmechanics.com/outdoors/survival/
tips/4220516

www.ready.gov

www.survivalblog.com

Recommended Reading

Aguirre, Fernando Ferfal. *Surviving the Economic Collapse.*
Buenos Aires, Argentina: Fernando Aguirre, 2009.

Barkdull, Larry. *Emergency Essentials Tips for Preparedness.*
Shadow Mountain, 2003.

Bartholomew, Mel. *Square Foot Gardening*. Emmaus, PA:
Rodale, 2005.

Better Homes and Gardens, *Canning and Preserving
Recipes.* Des Moines, IA: Meredith Publishing, 1996.

Dickson, Murray. *Where There Is No Dentist*. Hesperian
Foundation, 13th updated printing, 2010.

Freed, Dolly. *Possum Living*. Portland, OR: Tin House
Books, 2010.

Layton, Peggy. *Emergency Food Storage and Survival
Handbook*. Roseville, CA: Prima Publishing, 2002.

Lundin, Cody. *When All Hell Breaks Loose: Stuff You Need to
Survive When Disaster Strikes*. Gibbs Smith, 2007.

Maxwell, Jane, Carol Thuman, and David Werner. *Where
There Is No Doctor: A Village Health Care Handbook*.
Hesperian Foundation, revised edition, 1992.

Piven, Joshua. *The Complete Worst-Case Scenario Survival
Handbook, Man Skills*. San Francisco: Chronicle Books,
1999–2010.

Rombauer, Irma S., Rombauer Becker, Marion and Becker,
Ethan. *Joy of Cooking: All About Canning and Preserving*.
New York: Simon & Schuster, 2002.

Storey, John and Martha. *Storey's Basic Country Skills*.
North Adams, MA: Storey Publishing, 1999.

Stroud, Les. *Survive! Essential Skills and Tactics to Get You
Out of Anywhere Alive*. New York: Harper Collins,
2008.

Twitchell, Mary. *What to Do When the Power Fails*. North
Adams, MA: Storey Publishing, 1999.

Tymothy, Cy. *Sneaky Uses for Everyday Things*. Kansas City,
MO: Andrews McMeel Publishing, 2003.

U.S. Army Survival Manual FM 21-76.

Conversions

Measure	Equivalent	Metric
1 teaspoon	–	5 milliliters
1 tablespoon	3 teaspoons	14.8 milliliters
1 cup	16 tablespoons	236.8 milliliters
1 pint	2 cups	473.6 milliliters
1 quart	4 cups	947.2 milliliters
1 liter	4 cups + 3½ tablespoons	1000 milliliters
1 ounce (dry)	2 tablespoons	28.35 grams
1 pound	16 ounces	453.49 grams
2.21 pounds	35.3 ounces	1 kilogram
100°F / 200°F / 350°F	–	38°C / 93°C / 175°C

OTHER ULYSSES PRESS BOOKS

Bug Out: The Complete Plan for Escaping a Catastrophic Disaster Before It's Too Late
Scott B. Williams, $15.95

Cataclysmic events strike sleepy towns and major cities every year. Being prepared makes the difference between survival and disaster. Guiding you step by step, *Bug Out* shows you how to be ready at a second's notice.

Bug Out Vehicles and Shelters: Build and Outfit Your Life-Saving Escape
Scott B. Williams, $15.95

When disaster strikes, the wise few who have planned ahead will survive in style. This intricately detailed guide shows readers how to prepare vital supplies, determine a safe bivouac location, outfit an automobile for emergency transport and long-term occupancy, and build the ideal shelter for any family.

Clutter Rehab: 101 Tips and Tricks to Become an Organization Junkie and Love It!
Laura Wittmann, $12.95

Make even the most chaotic home a well-organized, relaxed environment. Featuring quick solutions and simple projects that have been tested and approved by the loyal readers of the author's popular blog, *Clutter Rehab* offers a plan to tackle clutter and tidy every room.

Getting Out Alive: 13 Deadly Scenarios and How Others Survived
Scott B. Williams, $14.95

A unique combination of fictional scenarios, true accounts, and instructive information, *Getting Out Alive* presents captivating stories of people stranded and fighting for their lives against harsh, unmerciful conditions.

Special Forces Survival Guide: Wilderness Survival Skills from the World's Most Elite Military Units
Chris McNab, $17.95
With detailed instructions, helpful photographs and step-by-step illustrations, this book arms readers with the same battle-tested techniques used by the military's bravest, most elite soldiers.

Ultimate Guide to Wilderness Living: Surviving with Nothing but Your Bare Hands and What You Find in the Woods
John McPherson and Geri McPherson, $15.95
Packed with in-depth instruction and photos, *Ultimate Guide to Wilderness Living* teaches you the skills need to survive and live in the wild. Learn techniques for everything from starting a fire, building a shelter, and finding food to weaving baskets, making pottery, and fashioning traps and other hunting equipment.

The U.S. Army Survival Manual: Department of the Army Field Manual 21-76
Headquarters, Department of the Army, $13.95
Drawing on centuries of training and field testing, this guide covers every imaginable scenario such as finding drinking water in the desert, stalking game in the arctic, building a fire in the jungle, and recognizing signs of land when lost at sea.

To order these books call 800-377-2542 or 510-601-8301, fax 510-601-8307, e-mail ulysses@ulyssespress.com, or write to Ulysses Press, P.O. Box 3440, Berkeley, CA 94703. All retail orders are shipped free of charge. California residents must include sales tax. Allow two to three weeks for delivery.

ACKNOWLEDGMENTS

To project manager Kelly Reed and managing editor Claire Chun, thank you for all your assistance throughout the book writing process. To acquisitions editor Keith Riegert, thanks for your interest in my work.

To Scott B. Williams, author of three great books which I recommend highly, thank you for giving me the opportunity to review your book, which started the whole process.

To the prepping community online, which gave me inspiration and encouragement to keep writing.

ABOUT THE AUTHOR

BERNIE CARR became fascinated with survival techniques and self-sufficiency as a child, hearing stories of her father's adventures in the wilds of Southeast Asia as a land surveyor and avid outdoorsman. As an adult, she developed an interest in emergency preparedness and self-reliance, having survived the 1991 Northridge earthquake in California, the 1992 Los Angeles riots, and the evacuation of her home during the Southern California wildfires. She relocated to Houston, Texas, in an effort to avoid more natural disasters only to arrive in time to encounter the fury of Hurricane Ike in 2008.

Bernie has a bachelor of science degree from the University of Southern California and has worked as a technical writer in various fields such as personal finance, insurance, and health care. She writes *The Apartment Prepper's Blog* at apartmentprepper.com and resides in Texas with her family, along with her two cats, Cesar and Cleo.